Springer Tracts in Nature-Inspired Computing

The book series is aimed at providing an exchange platform for researchers to summarize the latest research and developments related to nature-inspired computing in the most general sense. It includes analysis of nature-inspired algorithms and techniques, inspiration from natural and biological systems, computational mechanisms and models that imitate them in various fields, and the applications to solve real-world problems in different disciplines. The book series addresses the most recent innovations and developments in nature-inspired computation, algorithms, models and methods, implementation, tools, architectures, frameworks, structures, applications associated with bio-inspired methodologies and other relevant areas.

The book series covers the topics and fields of Nature-Inspired Computing, Bio-inspired Methods, Swarm Intelligence, Computational Intelligence, Evolutionary Computation, Nature-Inspired Algorithms, Neural Computing, Data Mining, Artificial Intelligence, Machine Learning, Theoretical Foundations and Analysis, and Multi-Agent Systems. In addition, case studies, implementation of methods and algorithms as well as applications in a diverse range of areas such as Bioinformatics, Big Data, Computer Science, Signal and Image Processing, Computer Vision, Biomedical and Health Science, Business Planning, Vehicle Routing and others are also an important part of this book series.

The series publishes monographs, edited volumes and selected proceedings.

Nilanjan Dey
Editor

Applications of Ant Colony Optimization and its Variants

Case Studies and New Developments

Springer

Editor
Nilanjan Dey
Department of Computer Science
and Engineering
Techno International New Town
New Town, West Bengal, India

ISSN 2524-552X ISSN 2524-5538 (electronic)
Springer Tracts in Nature-Inspired Computing
ISBN 978-981-99-7226-5 ISBN 978-981-99-7227-2 (eBook)
https://doi.org/10.1007/978-981-99-7227-2

This Springer imprint is published by the registered company Springer Nature Singapore Pte Ltd.
The registered company address is: 152 Beach Road, #21-01/04 Gateway East, Singapore 189721, Singapore

Paper in this product is recyclable.

Preface

Ant colony optimization (ACO) is a metaheuristic method inspired by ant foraging behavior. It employs the concept of pheromone trails to solve difficult optimization problems quickly by emulating the collective intelligence of ant colonies. This book highlights the concept and origin of ant colony optimization and its variants. It includes several optimization-based applications using ant colony optimization by covering different applications. The book includes eleven chapters, where Misra and Chakraborty in Chap. 1 aim to encourage scientists to use ant colony optimization (ACO) and its new variations to tackle difficult engineering design problems. It provides an overview of ACO's fundamental principles, parametric configuration for ACO, its different variations, and examples of ACO usage in different fields. Then, Pérez-Carabaza et al., in Chap. 2, focused on a popular variant of the rank-based ant system. The chapter tries to improve the rank-based ant system, which has shown promise in experimental evaluations over numerous instances of the traveling salesman (TSP) and sequential ordering (SOP) problems. This ACO algorithm included two innovative diversification techniques: pheromone smoothing and reinforcement of novelty. In Chap. 3, Hashemi and Dowlatshahi presented a survey on feature selection techniques based on ACO. In this study, the authors discussed the problem of feature selection and the importance of ACO-based feature selection methods in machine learning. They then classified these approaches according to class labels. Furthermore, in Chap. 4, an examination of several ant colony optimization algorithm techniques for the diagnosis of diabetes was provided by Moharana et al. In this study, rule-based categorization systems were successfully extracted from data mining systems using ACO. Additionally, a set of guidelines for diagnosing diabetes illness was derived using ACO. In Chap. 5, by utilizing the idea of opposed pheromones, Farahmand-Tabar and Shirgir introduced an algorithm that attempted to improve the optimization process. In this method, ants exhibit both exploitation and exploration behavior while following pheromone trails. However, in contrast to conventional ACO algorithms, this strategy added the idea of oppositional pheromones, which stand for divergent answers or directions.

In Chap. 6, Sahinkaya and Ustun used ant colony optimization (ACO) to build an S-box based on the 2D logistic-sine linked map. They employed a variant of ACO for continuous optimization in their work to optimize the parameters of 2DLSCM. In Chap. 7, Murugesan et al. used a proportional integral derivative regulator to manage the load frequency (LFC) of a triple equal-area thermal power system using renewable energy sources. Using an integral of time multiplied absolute error (ITAE) criterion, the ant colony optimization (ACO) algorithm design parameters, such as pheromone strength, ants, evaporation rate, and iteration, are modified for a three-area power system devoid of any renewable energy sources. The proposed method's results are compared to some recently published heuristic methods, such as cohort intelligence (CIO), genetic algorithms (GAs), teaching learning-based optimization (TLBO), differential evolution (DE), and PID controllers trained using particle swarm optimization (PSO), for the same networked power system. In Chap. 8, Farahmand-Tabar and Shirgir introduced the antlion-facing ant colony optimizer (AFACO), a robust hybrid algorithm. AFACO blends the excellent search capabilities of ant colony optimization (ACO) with the effective hunting abilities of antlions. AFACO navigates paths based on pheromone values on edges while maintaining the antlion's predation process by adopting the discrete movement approach of ants. In Chap. 9, Sreelaja and Sreeja developed a hierarchical heterogeneous ant colony optimization method. To model the IoT devices in a group, a binary tree representation is used. Every IoT device has a unique identifier. A device's Boolean member function determines the presence of an IoT device in the device group. In Chap. 10, Boopathi et al. discussed the use of the ant colony optimization (ACO) technique in a thermal power system to improve system frequency stability. An unconnected and connected thermal power network was investigated. In the final chapter, Farahmand-Tabar explored the frequency-based optimization of truss dome structures in the continuous domain utilizing upgraded ant colony optimization methods. The study's goal was to optimize the size or size/shape of the truss structure while keeping frequency limits in mind.

The editor is thankful to the outstanding authors and referees for their contributions and outcomes. Their diligence and collaboration led to this remarkable book. The appreciation is also directed to the members of the Springer team for their support.

New Town, India Nilanjan Dey

Contents

About the Editor

Nilanjan Dey is an Associate Professor, Department of Computer Science and Engineering, Techno International New Town, Kolkata, India. He is a visiting fellow of the University of Reading, UK. He was an honorary Visiting Scientist at Global Biomedical Technologies Inc., CA, USA (2012–2015). He was awarded his Ph.D. from Jadavpur University in 2015. He has authored/edited more than 70 books with Elsevier, Wiley, CRC Press and Springer, and published more than 300 papers. He is the Editor-in-Chief of International Journal of Ambient Computing and Intelligence, IGI Global, Associated Editor of International Journal of Information Technology, Springer. He is the Series Co-Editor of Springer Tracts in Nature-Inspired Computing, Springer, Series Co-Editor of Advances in Ubiquitous Sensing Applications for Healthcare, Elsevier, Series Editor of Computational Intelligence in Engineering Problem Solving and Intelligent Signal processing and data analysis, CRC. His main research interests include Medical Imaging, Machine learning, Computer Aided Diagnosis, Data Mining etc. He is the Fellow of IETE and Senior member of IEEE.

Chapter 1
Ant Colony Optimization—Recent Variants, Application and Perspectives

Bitan Misra and Sayan Chakraborty

1 Introduction

A population-based metaheuristic named ant colony optimization (ACO) can be used to approximate solutions to difficult optimization issues and was first presented by Dorigo et al. [1]. ACO is a type of metaheuristic class of approximate algorithms, which are applied to obtain workable solutions to challenging combinatorial optimization (CO) problems in an acceptable period of time. Swarm intelligence, which is a subset of bioinspired algorithms, involves the ant colony optimization approach. ACO is an artificial ant colony that relies on pheromone trails to promote indirect communication inside a colony of ants, simulating the biological process of ants seeking out food in the wild [2, 3].

Many researchers have focused on the topic of ACO, and there are now many study findings that are both experimental and theoretical in character. NP-hard combinatorial optimization issues were the focus of the first applications of ACO. Unsurprisingly, this field still contains the largest corpus of ACO research. Most of the studies on ACO that have been published thus far clearly focus on using it to solve computationally difficult problems [4]. Although most of this research focuses on academic implementations, it is significant that businesses have begun using ACO algorithms for practical purposes. Solutions to the 'travelling salesman' problem that are nearly ideal [5] have also been obtained using it. In situations where the graph may vary dynamically, they are superior to simulated annealing [6–9] and genetic algorithm [7–9] techniques because the ant colony algorithm may run constantly and adjust to modifications in real time. Gutjahr was the first to establish a theoretical

B. Misra (✉) · S. Chakraborty
Department of Computer Science and Engineering, Techno International New Town, Kolkata, India
e-mail: bitan.misra@tint.edu.in

S. Chakraborty
e-mail: sayan.chakraborty@tint.edu.in

N. Dey (ed.), *Applications of Ant Colony Optimization and its Variants*, Springer Tracts in Nature-Inspired Computing, https://doi.org/10.1007/978-981-99-7227-2_1

basis and demonstrate the ACO algorithm's probability of convergence [10]. Paper [11] provides a summary of the theoretical findings that are relevant to ACO. Arcelor and Mittal additionally used ACO algorithms to address a number of optimization issues that arise in their manufacturing facilities [12, 13]. In situations where the graph may vary dynamically, they are superior to simulated annealing and genetic algorithm techniques because the ACO algorithm may run continually and adjust to modifications in real time.

Regarding applications, the development of parallel implementations that can employ the new parallel hardware that is currently accessible, as well as the application of ACO algorithm for solving dynamic, multiobjective, stochastic, continuous and mixed-variable optimization issues, are both present-day trending issues. Features such as time-varying data, numerous objectives or the accessibility of stochastic knowledge about situations or data are quite typical in real-world scenarios. It is intriguing to observe that people are paying more attention to ACO applications in situations with these traits. In fact, ACO algorithms can be thought of as especially helpful in 'ill-structured' issues where local search is difficult to implement or in highly dynamic areas where solely local information is accessible.

2 Background

In ACO, a set of software agents referred to as artificial ants seek out effective solutions to a considered optimization issue. The optimization issue is modified into the job of finding out the optimum trail on a weighted graph to employ ACO. ACO algorithms are population-based metaheuristics that take their cue from ant foraging patterns. Many ant species are almost entirely blind in nature, and they communicate with one another using a chemical compound called pheromone. For the benefit of pursuing ants, ants emit pheromone, a chemical substance. Among the most effective channels for communicating, pheromone-dependent communication is often seen in the environment. Social insects like bees, ants and termites use pheromones for inter-agent and agent-swarm communication. Stigmergy is the name of this indirect communication technique. Artificial pheromones are applied in multirobot and swarm robotic systems due to their viability.

These pheromones are utilized by artificial ants to obtain the best course of action (route) for a variety of situations that are ambiguous and combinatorial in nature. When moving, ants leave pheromone deposits on the travelled tracks, creating pheromone trails. The subsequent ants select the track with a higher pheromone concentration after detecting its existence. These pheromone trails have a high degree of volatility. It will disappear after a certain period of time. Similar to real ants, artificial ants also have an evaporation factor that aids in deprioritizing nonoptimal pathways. One of the two possible initial pheromone quantities is 1 or 0. If the initial pheromone value is 1, the rate of evaporation will be set up so that the pheromone will first dissipate before covering the path. The first laying property will be created if it is set to 0, and the pheromone evaporation factor will then minimize the quantity of pheromone placed on the route. The pheromone evaporation rate describes the rate at which pheromones vanish over a period of time. In artificial ants, this evaporation is crucial for avoiding nonoptimal pathways and accelerating convergence. The term pheromone decay coefficient refers to the constant rate at which a pheromone degrades.

Ants initially move aimlessly around their nests in quest of food. This random search opens up a variety of pathways from the nest to the food source. Depending on the type and quantity of food, ants now bring back a portion of it together with the necessary pheromone concentration. Depending on these pheromone trials, the likelihood that succeeding ants would pick a specific path to the food source would act as a directing factor. Clearly, this likelihood depends on both the amount present and the rate of pheromone dissipation. The pheromone evaporation rate is also an influential variable, therefore it is simple to account for the length of each path.

The concept may be analysed by considering a simple example where there are only two routes available from the colony to the food location. Initially, we assume that all ants are in their initial position or nest. Now, the steps can be analysed as follows:

1. Initially, no pheromones are present in any of the paths. Therefore, the probability of an ant choosing any of the paths is 50%. From Fig. 1a, Path 1 and Path 2 are separated by a significant distance between them. Therefore, the ant that chooses the shortest path (Path 2) reaches the destination in a shorter amount of time, as shown in Fig. 1b.
2. It appears that they are now faced with a similar selection dilemma, but this time, picking the shorter path is more probable as the pheromone trail along it is already present. The early feeder ant now returns through the quickest route after finding food and bringing some food with it. As it moves along the returning path, it deposits pheromone on the ground. This is shown in Fig. 1c.

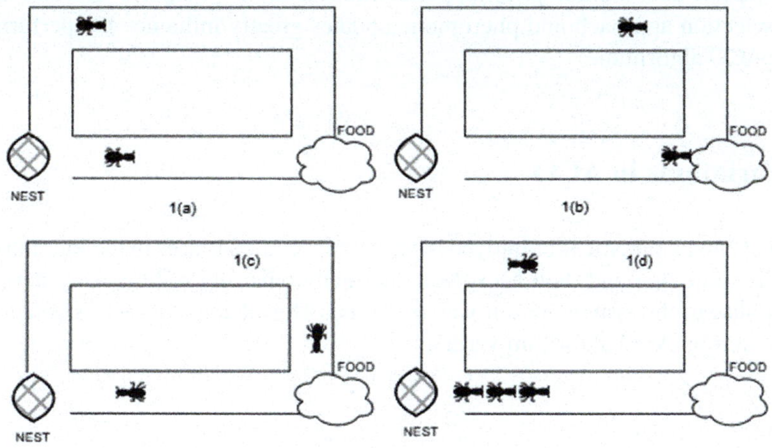

Fig. 1 Schematic diagram of ant colony optimization (ACO)

3. When the third ant makes the decision to travel in search of food, it will decide which route will cover the shortest length based on the quantity of pheromones present on that track. More pheromones are present on a shorter journey than on a longer journey. As seen in Fig. 1d, the third ant will choose the shorter route since it is practical.
4. The shorter route contains a more amount of pheromone than other paths and therefore the ants are more inclined to follow that path after multiple iterations of this process. In addition, evaporation reduces the pheromone quantity along the longer path, decreasing the likelihood that it will be taken in the future. The entire colony consequently uses the quicker route more frequently. Thus, path optimization is completed.

Now that we are aware of how the ants acted in the example above, we can construct an algorithm. In order to keep things simple, we have just considered one food source and one ant colony with n ants. The probability of each ant travelling from node i to node j at the t-th iteration is calculated as follows:

$$P_{i,j}(t) = \frac{\alpha_{i,j}}{\sum_{i-1}^{n} \alpha_{i,j}} \tag{1}$$

In Eq. (1), $\alpha_{i,j}$ defines the pheromone amount present in the route from i to j. The quantity of pheromone present in this path is calculated as follows:

$$\alpha_{i,j} = (1 - \rho)\alpha_{i,j} + \Delta\alpha_{i,j} \tag{2}$$

In Eq. (2), ρ defines the rate of disappearance pheromone and $\Delta\alpha_{i,j}$ represents the deposited pheromone in that path. The number of ants may impact the performance of ACO algorithm. The algorithm's chemotaxis operator also has an impact on performance and is crucial for determining where to discover food sources. Additional parameters, including initial population, iterations, distance measure metric, edge selection approach and pheromone updates greatly influence the performance of the ACO algorithm.

3 Variations in ACO

The ACO metaheuristic is employed in various specialized cases in the literature. Ant systems came first and then Max–Min and many other algorithms were developed in response to the issue where it was first used. This chapter discusses some of the most common ACO algorithm variants.

3.1 Ant System (AS)

The ant system can be considered as primary ACO algorithm. This algorithm is the same as the one that was previously discussed. Its main characteristic is that all ants that have finished the tour update the pheromone levels. The primary value of the basic ant system (AS) lies in its role as a model for multiple ant algorithms that together implement the ACO paradigm. Making artificial ants walk across graph nodes while having the method run t times, t being the number of iterations. The creation of the tour and the updating of the pheromones on the nodes that the ant visits are the two key processes in the ant system. Each ant will be initially placed in an apparently random area. A state transition rule (probabilistic action) will be provided to ant n at each phase of the tour-building process. The pheromone update will be made after all the ants have built the path. This will be carried out after a steady amount of pheromone strength on the entire arc has been reduced. The ant system's artificial ants exhibit comparable behaviours. They differ from their biological relatives in two ways. First, they are not blind; rather, they are aware of their surroundings and make use of this knowledge to be both avaricious and adaptive. Second, they also possess a memory, which is important to guarantee that only workable solutions are produced. Positive feedback, distributed computing and the application of a beneficial greedy heuristic are this model's key features [14].

The main benefit of the ant system is that new solutions are obtained with each iteration rather than the current ones being changed. The ant system is not dependent on neighbourhoods as a result. However, the ant system methodology still allows for the use of positive feedback within the population, which is a key benefit of population-based systems. The ant system is a distributed metaheuristic that continually creates solutions to challenging combinatorial optimization problems by combining a local heuristic feature with adaptive memory. In earlier research [15–20], it was discovered that the AS algorithm can effectively address a wide range of issues, including quadratic problems [16–18], graph colouring [19, 20], job shop scheduling [15–20] and vehicle routing [16, 18]. It displays even better worst-case performance than the other techniques and can compete with average behaviour [15].

3.2 Ant Colony System (ACS)

The ant colony system (ACS), first developed by Dorigo and Gambardella in 1997 [21], was the first significant upgrade to the original ant system. The first difference with the ant system is in terms of the decision rule and pseudorandom proportional rule. The second difference is in terms of local pheromone update. Finally, the difference lies in terms of the best only offline pheromone updates. An improved variant of the ant system (AS) algorithm is the ACS algorithm, which was developed to investigate the actions of actual ants. It replicates the method applied by ants to recall the paths from their nest to the food and this involves a group of artificial ants working together to search for the best solution by exchanging the information found in the pheromone deposits of the various trajectories. The algorithm has a significantly simpler structure. Additionally, it has fewer control parameters than many other approaches, which simplifies the process of configuring the parameters [22]. The pseudorandom proportional rule is utilized by ants in ACS. There is a wealth of material on vehicle routing issues. Most of the effort is focused on static problems, in which all of the data are known before the optimization process even begins. Dynamic vehicle routing challenges, where new demands must be dynamically accommodated into an expanding schedule as they are received over time, are a new class of issues brought on by recent technical advancements [23]. Based on an ACS method, article [24] discusses multiple ACS for vehicle routing problems with a time window algorithm that utilizes two ant colonies to reduce both the number of vehicles and the overall distance travelled. For the traditional VRP and its application to the time-dependent scenario, the multiple ACS is described in [25]. In [26], the major distribution system planning problem is resolved by combining the ACS methodology with a traditional load-flow algorithm.

The effectuation of the ACS algorithm can be enhanced by incorporating the idea of parallel processing. According to the experiments in [27], new exchange techniques depending on a weighting system enhance performance, specifically in tactics that distribute pheromone data throughout all colonies. An efficient method for obtaining a collection of nondominated solutions that concurrently reduces total resource waste and power dissipation can be solved efficiently by a multiobjective ant colony system algorithm [28]. Depending on a given objective function, an ACS-based virtual machine consolidation (ACS-VMC) method [29] locates a close to ideal solution. Experimental findings on actual Workload patterns show that ACS-VMC lowers energy consumption in cloud data centres while maintaining the required performance level. To implement virtual machine placement (VMP), an ACS-based strategy is devised in [30]. According to the literature, an ant colony optimization approach integrated with a local search strategy called a multiobjective ant colony system algorithm can provide solution to the NP-hard problems. Combinatorial optimization problems are better solved using the ant colony system (ACS) paradigm. It is possible to draw a comparison between how ants locate the shortest route between their source (the nest) and their destination (food) and how they solve the challenging combinatorial unit commitment problem (UCP) [31] to arrive at the minimal cost path (MCP), where parallel versions of the ACS deviate the most is in how the pheromone memory is implemented. The three revolutionary parallel ACSs for graphics processing units (GPUs) are briefly explored in [32]. The results show that, in the majority of cases, the produced solutions were of a much higher calibre than for the sequential ACS.

3.3 Max–Min Ant System (MMAS)

Another innovation is the MAX–MIN ant system (MMAS). This variant of ACO was put up by Stützle and Hoos in 2000 [33]. The Ant System was demonstrated to be a practical approach to tackling challenging combinatorial optimization challenges. However, it demonstrated poor accomplishment on several popular benchmark issues, including the travelling salesman issue. Therefore, introducing algorithmic changes to attain much better performance has been a key focus of research on ACO algorithms. In contrast to AS, MMAS specifically limits the minimum and maximum pheromone values, and only the best ants can add pheromone trails. As opposed to being expressly established by the algorithm designer, these values are implicitly restricted in AS and ACS, which means that the limits' value is determined by how well the algorithms perform. Iteration best or global best ants are the two types of ants that are permitted to update the pheromone. The only tour that is allowed to add pheromone to its route in MMAS is the global best tour or the best iteration tour. The pheromone's minimal value is often determined through experimentation; however, [33] has provided some theories on how to quantify its value quantitatively. If the ideal length of the ant tour is known, it is possible to analytically calculate the highest value of the pheromone. When the ant is going to choose an arc to travel, it will be prevented from doing so if the arc does not lie inside the interval. According to what was learned from the prior history of pheromone trails, that arc might not result in a satisfactory grade of solution, hence the deletion. To force a deeper examination of potential solutions, all edges are initialized to the maximum value of pheromone. However, if after a certain number of iterations, no improvement is visible, the process is redone. Hence, an algorithmic variation called MMAS is developed that performs better than AS to demonstrate how ACO methods could be effective replacement for current algorithms for solving challenging combinatorial optimization issues [33]. By enabling only the best solutions to deposit pheromone during the pheromone trail update, MMAS successfully exploits the search history. This is one advantage that the MMAS algorithm demonstrates over the basic ant system algorithm. Additionally, by using a relatively simple technique to regulate the intensities of the pheromone trails, early convergence of the search is effectively prevented. The MMAS can be easily expanded using local search algorithms, which are the last step. For the TSP and the Quadratic Assignment Problem (QAP), MMAS is now among the top ACO algorithms, according to the authors in article [33].

Further applications of MMAS are made to symmetric and asymmetric travelling salesman issues [34]. Additionally, it is demonstrated that incorporating a local search phase to MMAS can considerably improve its performance while also enhancing the ant-generated solutions [35]. Additionally, the solution quality for asymmetric TSP obtained with MMAS was significantly better than that for the genetic algorithm. Using the weighted sum method, MMAS exhibits appreciable efficiency in multi-objective optimization problems. Additionally, the solution quality for asymmetric

TSP obtained with MMAS was significantly better than that for the genetic algorithm. Using the weighted sum method, MMAS exhibits appreciable efficiency in multiobjective optimization problems.

3.4 Elitist Ant System (EAS)

Even though this trail has not been retraced, the global best solution, similar to all the other ants in the elitist ant system (EAS), leaves pheromone behind after each iteration. The commonly encountered problem of vehicle route allocation can be successfully solved by using the EAS algorithm [36]. In various types of optimization problems, the EAS algorithm surpasses the GA and ACS in terms of function value and constraint correctness. In [37], the EAS approach is employed to maintain a trade-off between cost and weight for the objective functions. Increasing pheromone on the edges of the best path (i.e. the shortest path) and dropping pheromone on the edges of the worst route (i.e. the longest route) are two features of the hybrid EAS technique introduced in [38] that only uses global updates. The vehicle routing problem (VRP), a popular NP-hard problem, is successfully fixed using hybrid EAS [38]. The hybrid elitist ant system technique is recognized as a workable population-based solution for a few optimization problems.

A hybrid elite ant strategy is employed in [39] to make use of the clustering problem's search space. The findings of six benchmark datasets for the medical clustering problem using an elitist ant system strategy in comparison to results from other methodologies serve as evidence of its potential. The suggested adaptive elitist ant system methodology can, according to computational results, offer solutions of a higher quality than the hybrid elitist ant system technique and other methods in the literature. Nevertheless, To assess how well the EAS algorithm performs when used to solve the scattered job shop scheduling problem, numerous numerical experiments are carried out [40], and the results demonstrate the EAS algorithm's limitations when compared to other methods from the literature. An example of a hybrid EAS application is the travelling salesman problem [41], where people can find the shortest routes using information from pheromone trails and heuristics. The findings demonstrate that broadening the search while placing a focus on superior solutions is effective, efficient and reliable. This exceptional achievement is attributed to the application of an elite pool together with intensification and diversification procedures. The external memory layout of a hybrid EAS is referred to as an elite pool and contains high-quality and diversified solutions to equalize the quality and diversity of the search. EAS maintains the quality of the solutions while balancing search diversification and intensification. To attain sufficient dominance over search exploration and exploitation, this approach uses variety and intensity to assist with elite pheromone update and pheromone evaporation. The diversification technique prevents early convergence, while the intensification technique is employed to study a solution's neighbours more thoroughly. Examining earlier experiments [42] demonstrates that EAS has found superior solutions by utilizing guided search exploration and problem-specific knowledge exploitation.

3.5 Rank-Based Ant System

One illustration that expands the basic AS is the rank-based ant system. The Bullnheimer [43] version of the ant system that integrates ranking into the pheromone updating procedure is known as the rank-based Ant System. All solutions are ranked in this algorithm based on their length. In this iteration, only a predetermined number of the top ants are permitted to upgrade their experiments. For each solution, the quantity of pheromone accumulated is weighted, resulting in shorter solutions depositing more pheromone than longer solutions. By placing pheromones solely on the edges that are travelled by top-ranked ants, this function increases the convergence of the system. As a result, by concentrating pheromones on particular edges, the system converges more quickly. However, the system usually settles into local solutions since pheromones frequently focus only on particular edges [43].

This model resolves the TSP using the pheromone method update for simulation exploration. The study in [44] solves dynamic TSPs using rank-based ACO. The given rank-based ACO method efficiently examines paths that are impacted by a change in the environment by adding the rank-based nonlinear selective pressure function and the updated Q-learning model into the original ACO. A rank-based ant colony optimization method improves the convergence properties of the original ACO, which uses a random proportional rule to define the likelihood of an ant colony exploring a city. This random proportional rule's probability distribution, that is comparable to that of the stochastic universal sampling technique frequently used in genetic algorithm selection, is effective for advantageous paths in small TSPs but ineffective for obtaining such paths in large TSPs. An efficient rank-based ant system algorithm can be considered a solution because this problem falls under the category of NP-hard problems [45]. This algorithm uses techniques to slow convergence and promote exploration while reinforcing local search in the vicinity of the best solution identified in each iteration. Hard combinational optimization issues have been proven to respond well to the rank-based ant system algorithm [46]. The program effectively resolves a discrete design variable weight minimization problem for structures [47]. The algorithm, however, incorporates actions that belong to a local solution and are absent from the real biological system. By permitting agents to reflect their place in a local hierarchy, the rank-based ant colony algorithm in [48] is able to achieve even more improvements.

3.6 Parallel Ant Colony Optimization (PACO)

In the past 10 years, parallel implementations have gained popularity as a way to increase the effectiveness of population-based metaheuristics. Even when dealing with challenging optimization problems, parallel execution of metaheuristics enables achieving good quality solutions in an acceptable amount of time by dividing the population into multiple processing components. The relevant literature contains numerous publications that claim parallel implementations can enhance the ACO exploration pattern. Given a set amount of time to search, one of the numerous benefits of parallel ant colony optimization (PACO) is that it improves the quality of the solutions identified Additionally, given a predetermined solution quality, the goal is to shorten the time required to obtain a solution that is not poorer than that quality [49]. The original parallel ACO concepts concentrated on conventional supercomputers and workstation clusters. Modern multicore processors, graphics processing units (GPUs), and grid computing environments offer new potential to utilize parallel computing approaches to enhance ACO performance and shorten computation times [50]. Numerous publications have concentrated on presenting experimental findings based on the resolution of popular combinatorial optimization issues, including TSP, QAP, MAXSAT and other well-known issues. These studies have primarily focused on demonstrating how parallel ACO models outperform conventional sequential implementations in terms of computing efficiency measures and/or result quality [49, 50].

There are no standardized taxonomies for categorizing the parallelization methodologies used with ACO, despite the research community's various parallel ACO implementation proposals. Parallel ant colony optimization for TSPs using a dynamic method on GPUs [51]. The quadratic assignment problem (QAP) can be solved by applying a parallel ant colony model [52]. A pheromone matrix that functions as a global memory enables collaboration between the artificial ants. To tackle an NP-hard issue, many computational resources are often needed. Particularly, as problem sizes expand, run-time overheads may become prohibitively expensive. By utilizing many processors in independent or shared memory architectures, parallel techniques offer a mechanism to handle such run-time problems [53].

3.7 Recursive Ant Colony Optimization (RACO)

The optimal answer for many geophysical problems can be found using the recursive ant colony system (RACO), an improved version of the ACO algorithm. As an extension of ACO, RACO optimizes an ant's route from the nest to the food supply by simulating their social behaviour. Comparable to the divide and conquer strategy-based recursive merger sort, the RACO algorithm employs a partitioning technique to the problem. It is a recursive adaptation of the ant system that divides the entire search domain into various subdomains and accomplishes the goal on each of them

[54, 55]. The iteration is carried out recursively in each subdomain to obtain the shortest route in that subproblem. The top few of all the subdomains are elevated to the next level when the results from all of them are compared. The process is continued until the output has the appropriate precision. The subdomains related to the chosen results are again divided. The technique is based on the observation that smaller problems with fewer cities have superior Ant Colony application efficiency [55]. This happens because of the algorithm's randomness; numerous good random judgements based on weighed choices must be combined to create an efficient solution, and as the problem's complexity grows, so does the number of decisions needed to produce a single tour. Edge Detection and Feature Extraction and the Travelling Salesman Problem are both solved using this method [54, 56].

4 Applications

The initial applications of ACO were centred on NP-hard combinatorial optimization problems. Unsurprisingly, the majority of ACO research is still concentrated on this area. Ant colony optimization algorithms are employed to resolve several combinatorial optimization issues, including vehicle routing, quadratic assignment and protein folding. Many of the resulting approaches have been used to solve stochastic problems, multitarget implementations and parallel implementations of dynamic problems involving real variables. The ACO method is also used to obtain solution to the TSP to provide outcomes that are very close to perfect. The present focus of ACO algorithm research is the improvement of theoretical concepts and the actual implementation of the metaheuristic to innovative, challenging problems. Since the ant colony algorithm may run continually and can adapt to modifications in real time, it has an advantage over techniques like simulated annealing and genetic algorithms when the graph may fluctuate dynamically. Both network routing and urban transit systems are interested in this. The application of ACO to resolve dynamic, multiobjective, stochastic, continuous and mixed-variable optimization issues is an increasingly popular topic in terms of applications. Additionally, by developing parallel implementations of the algorithm, the new parallel technology that is now available can be utilized by ACO method.

The travelling salesman problem is successfully solved using ant colony optimization approaches, for instance. To increase the solution quality, the travelling salesman problem (TSP) is addressed in this research using population-based ant colony optimization techniques and novel hierarchical pheromone updating. The travelling salesman problem, in which the objective is to determine the quickest round-trip between a number of cities, was solved using the world's first ACO algorithm, known as the ant system. The concept is essentially based on ants, who travel independently on one of the numerous round trips and visit numerous towns and cities. TSP is NP-hard because there is no 'quick' approach, and the challenge of finding out the best route increases as the number of destinations increases.

The first problem to which AS was applied was a problem with vehicle routing (VRPs), which is a logical extension of the TSP. In these situations, a group of automobiles stationed at a depot have to assist a group of customers before returning to the depot; the objective is to use the smallest number of vehicles while minimizing the overall distance travelled by each vehicle. Any combination of customers may be chosen throughout the process of choosing vehicle routes to determine the delivery path for each vehicle. The number of alternative solutions to the vehicle routing problem expands exponentially with the number of clients to be served, making it a combinatorial optimization challenge. The vehicle routing problem is regarded as NP-hard since there isn't a polynomial algorithm that can consistently identify the best answer. In the VRP, the rank-based ACO algorithm exhibits outstanding efficiency. It may be possible to gain further knowledge on when to employ a multiple colony technique to solve vehicle routing problems by testing larger problem sizes and cars with varying capabilities. One of the most challenging ACO problems, QAP can only be optimally handled in tiny occurrences. In fact, many real-world applications can be defined as QAPs, including scheduling, designing typewriter keyboards, and wiring backboards. Starting with AS and moving on to numerous of the more advanced versions, several ACO algorithms addressed the QAP. Another typical scenario for evaluating the effectiveness of ACO algorithms is scheduling issues.

In the past few years, ACO techniques started to be developed for issues that weren't typical combinatorial optimization [57, 58] testbeds but were instead more closely related to real-world scenarios, demonstrating the maturity of the area. Ant colony optimization (ACO) is a method with great potential for enhancing image edge recognition. ACO employed a vector known as pheromone sensitivity level, or PSL, to improve the accuracy and consistency of picture edge recognition. The ACO method is still a significant development in the field of CT scan image analysis research and can be used for the detection and segmentation of lung nodules with irregular shapes. The computation of high-dimensional information, where the overall amount of features may be unlimited or indefinite but the number of data examples is fixed, is the subject of a growing field called streaming feature selection. The ACO method may be employed to choose the features [59] that will allow a machine learning model to be trained to accurately detect cervical cancer. In the field of medicine, the ACO algorithm has also been used for classification and prediction. An ant colony optimization [60] algorithm for interpreting Bayesian classifiers combination is used to solve problems like heart disease and cardiotomy-based predictions [61], where it's important to be able to comprehend the results in order to make the right clinical judgements. ACO can be used to automatically detect exudates, a diabetic retinopathy lesion [62], in fundus pictures, aiding in the early detection of the condition.

Due to its many benefits, including intrinsic parallelism and positive feedback, which accelerates the identification of effective solutions, ACO is being used more and more in numerous industries. With the aid of ACO approaches, even more theoretical issues pertaining to geographical data analysis were addressed. Telecommunication is a recent and fascinating area of ACO study that is not immediately related to combinatorial optimization. In fact, it appears that the field of packet switching communications is a promising one for ACO-related routing techniques. As a reason for this, ACO can be vastly used in dynamic applications. However, uncertainty lies in terms of convergence time. Additionally, the probability distribution changes in each iteration, which is another drawback of this algorithm.

5 Future Scope

ACO is regarded as a stochastic probabilistic search that offers numerous study directions on NP-hard problems. Scheduling, routing, clustering, and many more optimization issues have seen success with the approach. There is a good chance that future study will produce remarkable findings that might one day offer solutions to combinatorial problems that are currently intractable. Through ACO, more possibilities for dynamically selecting the ideal location can be created. The application of ACO in the biomedical field is still developing and requires more research. Future research could focus on parallel search execution and space exploration, effective heuristic use, the pheromone evaporation factor, initial candidate solutions and initial ant numbers chosen in a way that prevents premature convergence.

Currently, single-objective optimization problems are the major application for the ACO method. There is scope to expand the technique to handle multiobjective optimization issues where there are several simultaneous objectives to optimize. To make ACO more effective at solving particular optimization issues, researchers can experiment by hybridizing it with other algorithms. ACO can be combined with genetic algorithms, tabu search and simulated annealing to provide better outcomes and speed up convergence. An enormous challenge in optimization is dealing with dynamic optimization problems, where the variables of the problem change over time. To successfully address such issues, future research can concentrate on introducing dynamic mechanisms into the ACO algorithm. ACO has been successfully used in solving minor and medium-sized issues. Its performance in complex issues is still problematic. The development of parallel and distributed ACO algorithms can manage complex problems of research. Finally, real-time applications such as network routing, traffic flow control and robotics can all employ ACO. ACO algorithms and real-time data can be combined by researchers to produce effective and efficient solutions to problems. Overall, the potential for expanding the application of ACO across a variety of optimization challenges is broad and exciting.

6 Conclusions

Designing efficient combinatorial optimization solution algorithms has historically and currently benefited from the ant colony optimization approach. ACO is one of the most successful methodologies in the metaheuristic field since both its application efficacy and its theoretical foundation have been proven after more than ten years of research. The travelling salesman problem (TSP), dynamic travelling salesman problem (DTSP), quadratic assignment problem (QAP) and other challenging combinatorial problems may all be handled using the ACO metaheuristic. Under specific circumstances, the ACO algorithm has been shown to converge to the best solution, and a number of enhancements have been suggested to boost its accuracy and productivity. The ACO algorithm has also been used in practical scenarios, including e-commerce platforms, sensor cloud technology and robotic motion planning, among others. ACO, however, is sensitive to parameter selection and initialization, thus it might need some fine-tuning to produce the desired outcomes. ACO can be computationally costly, particularly for complex situations. ACO does well at solving discrete problems, but because high-dimensional search spaces are more difficult to search in, it may do less well there. Instead of examining the whole search space, ACO may converge to local optima, producing less-than-ideal results. It is crucial to understand that despite its drawbacks, ACO may still be an effective algorithm in some situations. Its restrictions may change based on the problem or application. Overall, the ACO algorithm is a useful tool for decision-making and optimization, and it has a wide range of possible applications in the future across various research and technical industries.

References

1. M. Dorigo, Optimization, learning and natural algorithms (in Italian), Ph.D. Thesis, Dipartimento di Elettronica, Politecnico di Milano, Italy, 1992.
2. Dorigo M, Stützle T (2019) Ant colony optimization: overview and recent advances . Springer International Publishing, pp 311–351
3. Acan A (2004) An external memory implementation in ant colony optimization. In: International workshop on ant colony optimization and swarm intelligence. Springer, Berlin, Heidelberg, pp 73–82. (Sept 2004)
4. Alexandrov D, Kochetov Y (2000) Behavior of the ant colony algorithm for the set covering problem. In: Operations research proceedings 1999 selected papers of the symposium on operations research (SOR'99), Magdeburg, September 1–3, 1999, Springer, Berlin, Heidelberg, pp 255–260
5. Cook WJ, Applegate DL, Bixby RE, Chvatal V (2011) The traveling salesman problem: a computational study. Princeton University Press
6. Stodola P, Michenka K, Nohel J, Rybanský M (2020) Hybrid algorithm based on ant colony optimization and simulated annealing applied to the dynamic traveling salesman problem. Entropy 22(8):884
7. Hoseini P, Shayesteh MG (2013) Efficient contrast enhancement of images using hybrid ant colony optimisation, genetic algorithm, and simulated annealing. Digital Signal Process 23(3):879–893
8. Hoseini P, Shayesteh MG (2010) Hybrid ant colony optimization, genetic algorithm, and simulated annealing for image contrast enhancement. In: IEEE congress on evolutionary computation. IEEE, pp 1–6. (2010, July)
9. Coleman CM, Rothwell EJ, Ross JE (2004) Investigation of simulated annealing, ant-colony optimization, and genetic algorithms for self-structuring antennas. IEEE Trans Antennas Propag 52(4):1007–1014
10. Gutjahr WJ (2000) A graph-based ant system and its convergence. Futur Gener Comput Syst 16(8):873–888

11. Dorigo M, Blum C (2005) Ant colony optimization theory: a survey. Theoret Comput Sci 344(2–3):243–278
12. Díaz D, Valledor P, Areces P, Rodil J, Suárez M (2014) An ACO algorithm to solve an extended cutting stock problem for scrap minimization in a bar mill. In: Swarm intelligence: 9th international conference, ANTS 2014, Brussels, Belgium, September 10–12, 2014. Proceedings, vol 9. Springer International Publishing, pp 13–24
13. Fernandez S, Alvarez S, Díaz D, Iglesias M, Ena B (2014) Scheduling a galvanizing line by ant colony optimization. In: International conference on swarm intelligence. Springer International Publishing, Cham, pp 146–157. (Sept 2014)
14. Dorigo M, Maniezzo V, Colorni A (1996) Ant system: optimization by a colony of cooperating agents. IEEE Trans Syst Man Cybern Part B (Cybern) 26(1):29–41
15. Bullnheimer B, Hartl RF, Strauss C (1997) A new rank based version of the ant system: a computational study. Working Paper No. 1, SFB adaptive information systems and modelling in economics and management science, Vienna, 1997, to appear in CEJOR
16. Bullnheimer B, Hartl RF, Strauss C (1998) Applying the ant system to the vehicle routing problem. In: Voss S, Martello S, Osman IH, Roucairol C (eds) Meta-Heuristics: advances and trends in local search paradigms for optimization. Kluwer, Boston
17. Bullnheimer B, Kotsis G, Strauss C (1998) Parallelization strategies for the ant system. In: De Leone R, Murli A, Pardalos PM, Toraldo G (eds) High performance algorithms and software in nonlinear optimization. Kluwer Academic, Dordrecht
18. Bullnheimer B, Hartl R, Strauss C (1999) An improved ant system algorithm for the vehicle routing problem. Ann Oper Res 89:319–328. https://doi.org/10.1023/A:1018940026670
19. Costa D, Hertz A (1997) Ants can colour graphs. J Oper Res Soc 48:295–305
20. Colorni A, Dorigo M, Maniezzo V, Trubian M (1994) Ant system for job-shop scheduling. JORBEL-Belg J Oper Res Stat Comput Sci 34(1):39–53
21. Dorigo M, Gambardella LM (1997) Ant colony system: a cooperative learning approach to the traveling salesman problem. IEEE Trans Evol Comput 1(1):53–66
22. Pourtakdoust SH, Nobahari H (2004, September). An extension of ant colony system to continuous optimization problems. In International Workshop on Ant Colony Optimization and Swarm Intelligence (pp. 294–301). Berlin, Heidelberg: Springer Berlin Heidelberg. (Sept 2004)
23. Montemanni R, Gambardella LM, Rizzoli AE, Donati AV (2005) Ant colony system for a dynamic vehicle routing problem. J Comb Optim 10:327–343
24. Barán B, Schaerer M (2003). A multiobjective ant colony system for vehicle routing probiem with time windows. In: Applied informatics, pp 97–102. (Feb 2003)
25. Donati AV, Montemanni R, Casagrande N, Rizzoli AE, Gambardella LM (2008) Time dependent vehicle routing problem with a multi ant colony system. Eur J Oper Res 185(3):1174–1191
26. Gomez JF, Khodr HM, De Oliveira PM, Ocque L, Yusta JM, Villasana R, Urdaneta AJ (2004) Ant colony system algorithm for the planning of primary distribution circuits. IEEE Trans Power Syst 19(2):996–1004
27. Ellabib I, Calamai P, Basir O (2007) Exchange strategies for multiple ant colony system. Inf Sci 177(5):1248–1264
28. Gao Y, Guan H, Qi Z, Hou Y, Liu L (2013) A multi-objective ant colony system algorithm for virtual machine placement in cloud computing. J Comput Syst Sci 79(8):1230–1242
29. Farahnakian F, Ashraf A, Pahikkala T, Liljeberg P, Plosila J, Porres I, Tenhunen H (2014) Using ant colony system to consolidate VMs for green cloud computing. IEEE Trans Serv Comput 8(2):187–198
30. Liu XF, Zhan ZH, Deng JD, Li Y, Gu T, Zhang J (2016) An energy efficient ant colony system for virtual machine placement in cloud computing. IEEE Trans Evol Comput 22(1):113–128
31. Simon SP, Padhy NP, Anand RS (2006) An ant colony system approach for unit commitment problem. Int J Electr Power Energy Syst 28(5):315–323
32. Skinderowicz R (2016) The GPU-based parallel ant colony system. J Parallel Distrib Comput 98:48–60

33. Stützle T, Hoos HH (2000) MAX–MIN ant system. Futur Gener Comput Syst 16(8):889–914
34. Stützle T, Hoos H (1998) Improvements on the ant-system: introducing the MAX-MIN ant system. In: Artificial neural nets and genetic algorithms. Springer, Vienna. https://doi.org/10.1007/978-3-7091-6492-1_54
35. Stützle T, Hoos HH (1996) Improving the ant system: a detailed report on the MAX–MIN ant system. FG Intellektik, FB Informatik, TU Darmstadt, Germany, Technical Report AIDA–96–12
36. Negulescu SC, Oprean C, Kifor CV, Carabulea I (2008) Elitist ant system for route allocation problem. In: Proceedings of the 8th conference on applied informatics and communications. World Scientific and Engineering Academy and Society (WSEAS), Greece, pp 62–67. (Aug 2008)
37. Hemmatian H, Fereidoon A, Sadollah A, Bahreininejad A (2013) Optimization of laminate stacking sequence for minimizing weight and cost using elitist ant system optimization. Adv Eng Softw 57:8–18
38. Yousefikhoshbakht M, Didehvar F, Rahmati F (2014) An efficient solution for the vrp by using a hybrid elite ant system. Int J Comput Commun Control 9(3):340–347
39. Abuhamdah AF (2020) Adaptive elitist-ant system for medical clustering problem. J King Saud Univ Comput Inf Sci 32(6):709–717
40. Chaouch I, Driss OB, Ghedira K (2017) Elitist ant system for the distributed job shop scheduling problem. In: Advances in artificial intelligence: from theory to practice: 30th international conference on industrial engineering and other applications of applied intelligent systems, IEA/AIE 2017, Arras, France, June 27–30, 2017, Proceedings, Part I 30. Springer International Publishing, pp 112–117
41. Jaradat GM (2018) Hybrid elitist-ant system for a symmetric traveling salesman problem: case of Jordan. Neural Comput Appl 29:565–578
42. Jaradat GM, Ayob M (2010). An elitist-ant system for solving the post-enrolment course timetabling problem. In: International conference on bio-science and bio-technology. Springer, Berlin, Heidelberg, pp 167–176. (Dec 2010)
43. Bullnheimer B (1999) A new rank based version of the ant system: a computational study. Central Eur J Oper Res Econ 7:25–38
44. Liu JL (2005) Rank-based ant colony optimization applied to dynamic traveling salesman problems. Eng Optim 37(8):831–847
45. Yousefikhoshbakht M, Didehvar F, Rahmati F (2016) An effective rank based ant system algorithm for solving the balanced vehicle routing problem. Int J Ind Eng 23(1)
46. Vlachos ARISTIDIS (2013) Rank-based ant colony algorithm for a thermal generator maintenance scheduling problem. WSEAS Trans Circuits Syst 12(9):273–285
47. Capriles PV, Fonseca LG, Barbosa HJ, Lemonge AC (2007) Rank-based ant colony algorithms for truss weight minimization with discrete variables. Commun Numer Methods Eng 23(6):553–575
48. Sakiyama T, Uneme K, Arizono I (2021) Rank-based ant system via the relative position in a local hierarchy. Complexity 2021:1–6
49. Pedemonte M, Nesmachnow S, Cancela H (2011) A survey on parallel ant colony optimization. Appl Soft Comput 11(8):5181–5197
50. Manfrin M, Birattari M, Stützle T, Dorigo M (2006) Parallel ant colony optimization for the traveling salesman problem. In: Ant colony optimization and swarm intelligence: 5th international workshop, ANTS 2006, Brussels, Belgium, September 4–7, 2006. Proceedings, vol 5. Springer, Berlin, Heidelberg, pp 224–234
51. Zhou Y, He F, Qiu Y (2017) Dynamic strategy based parallel ant colony optimization on GPUs for TSPs. Sci China Inf Sci 60:1–3
52. Talbi EG, Roux O, Fonlupt C, Robillard D (2001) Parallel ant colonies for the quadratic assignment problem. Futur Gener Comput Syst 17(4):441–449
53. Thiruvady D, Ernst AT, Singh G (2016) Parallel ant colony optimization for resource constrained job scheduling. Ann Oper Res 242:355–372

54. Gupta DK, Arora Y, Singh UK, Gupta JP (2012) Recursive ant colony optimization for estimation of parameters of a function. In 2012 1st international conference on recent advances in information technology (RAIT). IEEE, pp 448–454. (Mar 2012)
55. Gupta DK, Gupta JP, Arora Y, Shankar U (2013) Recursive ant colony optimization: a new technique for the estimation of function parameters from geophysical field data. Near Surf Geophys 11(3):325–340
56. Huadong Z, Chaofan L, Nan J (2019) A path planning method of robot arm obstacle avoidance based on dynamic recursive ant colony algorithm. In: 2019 IEEE international conference on power, intelligent computing and systems (ICPICS). IEEE, pp 549–552. (July 2019)
57. Dey N, Ashour AS, Bhattacharyya S (eds) (2020) Applied nature-inspired computing: algorithms and case studies. Springer, Singapore
58. De D, Mukherjee A, Das SK, Dey N (eds) (2020) Nature inspired computing for wireless sensor networks. Springer, Singapore
59. Kavitha R, Jothi DK, Saravanan K, Swain MP., Gonzáles JLA, Bhardwaj RJ, Adomako E (2023) Ant colony optimization-enabled CNN deep learning technique for accurate detection of cervical cancer. In: BioMed Research International
60. Vimal S, Khari M, Crespo RG, Kalaivani L, Dey N, Kaliappan M (2020) Energy enhancement using multiobjective ant colony optimization with double Q learning algorithm for IoT based cognitive radio networks. Comput Commun 154:481–490
61. Bouktif S, Hanna EM, Zaki N, Khousa EA (2014) Ant colony optimization algorithm for interpretable Bayesian classifiers combination: application to medical predictions. PLoS ONE 9(2):e86456
62. Pereira C, Gonçalves L, Ferreira M (2015) Exudate segmentation in fundus images using an ant colony optimization approach. Inf Sci 296:14–24

Chapter 2
Advancements in Rank-Based Ant System: Enhancements for Improved Solution Quality in Combinatorial Optimization

Sara Pérez-Carabaza, Akemi Gálvez, and Andrés Iglesias

1 Introduction

Metaheuristics have become indispensable tools for addressing complex combinatorial optimization problems, particularly when deriving optimal solutions through exhaustive methods is unfeasible [33]. Well-performing and very popular metaheuristics currently described in the literature include genetic algorithms [15, 16], tabu search [12, 13], simulated annealing [19], particle swarm optimization [17, 32], differential evolution [27, 35], and different evolutionary computation and swarm intelligence methods [2, 10, 18]. Among them, Ant Colony Optimization (ACO) has come up as a popular choice that has been widely applied across a broad range of fields. Inspired by the path-finding behaviors of ants, this metaheuristic combines problem-specific heuristics with information learned in previous iterations to adaptively navigate the solution space, constantly refining paths based on the knowledge acquired from past iterations and preserved encoded in the pheromone trails.

Since the appearance of Ant System (AS) [21], the original ACO algorithm, a number of ACO variants have been introduced. Among the most notable of these variants are the Rank-based Ant System (AS_{Rank}, [3]), Max-Min Ant System (MMAS, [36]), and Ant Colony System (ACS, [7]). These variants, although still adhering to the foundational principles of the AS, establish more effective balances between exploration and exploitation through unique pheromone update or transition rules, thereby

S. Pérez-Carabaza · A. Gálvez · A. Iglesias (✉)
Department of Applied Mathematics and Computational Sciences, University of Cantabria, Avenida de los Castros s/n, 39005 Santander, Spain
e-mail: iglesias@unican.es

S. Pérez-Carabaza
e-mail: sara.perezcarabaza@unican.es

A. Gálvez
e-mail: galveza@unican.es

N. Dey (ed.), *Applications of Ant Colony Optimization and its Variants*, Springer Tracts in Nature-Inspired Computing, https://doi.org/10.1007/978-981-99-7227-2_2

yielding better results. Despite ongoing research in improving ACO metaheuristics, MMAS and ACS are broadly regarded as the best-performing variants according to several sources [1, 4, 8]. For example, [11] proposes an approach within this research line to diversify the solution space by pairing searching ants.

Among these recent advances, we bring into focus our enhancement of the rank-based AS [25] which has demonstrated promising results in experimental analyses over several instances of two popular problems: Traveling Salesman (TSP) and Sequential Ordering (SOP). This ACO algorithm incorporates two novel diversification strategies: (1) reinforcement or originality and (2) pheromone smoothing. The motivation behind these two diversification strategies is to improve the long-term results of AS_{Rank}. Despite its ability to rapidly obtain reasonably good solutions, its long-term outcomes are generally inferior to those achieved by other AS extensions, often leading to stagnation [8]. The reinforcement of originality strategy rewards innovative solutions, promoting the exploration of search space, while the pheromone smoothing strategy prevents early convergence into local minima and stagnation.

In this chapter, beyond situating our proposed algorithm within the context of existing ACO algorithms and providing a detailed description of it, we conduct a comprehensive analysis of two distinct combinatorial optimization problems: the TSP [23] and the Minimum Time Search (MTS) problem [20, 24]. The former, a classic problem that has historically drawn significant attention from researchers, consists of finding the shortest closed path that visits every city in a set only once. The MTS problem, on the other hand, presents a more intricate formulation that aims to optimize the flying routes of one or several Unmanned Aerial Vehicles (UAVs) by minimizing the time required to locate a lost target within an uncertain environment. MTS has relevant applications ranging from searching for survivors of natural disasters, such as fires or earthquakes, to locating military targets. By subjecting our algorithm to these diverse optimization problems, this chapter demonstrates its robustness and suitability for addressing a wide range of combinatorial optimization tasks.

This chapter is arranged as follows: Sect. 2 provides an overview of ACO algorithms and their most relevant variants. Section 3 describes our extension of the rank-based ant system with the two new strategies indicated above. Section 4 presents an in-depth analysis of the proposed ACO algorithm over two different combinatorial optimization problems: TSP and MTS. Lastly, Sect. 5 closes the chapter with the main conclusions. It also discusses potential avenues for future work.

2 ACO Algorithms

ACO algorithms are metaheuristics inspired by the foraging behavior of colonies of ants. They have gained popularity thanks to their effectiveness in addressing combinatorial optimization problems. This section provides an overview of the most relevant ACO algorithms described in the literature.

Algorithm 1 ACO metaheuristic

Require: ACO parameters
  $\tau \leftarrow$ initialise pheromone trails
  while termination condition not met **do**
    for $k = 1$ to m **do** ▹ for every ant
      for every step until the k-th ant has completed the tour **do**
        select node j to visit next according to the transition rule.
      end for
    end for
    Daemon actions (optional)
    Update pheromone trails τ
  end while

Fig. 1 General scheme of an ACO algorithm

ACO algorithms involve a given population of artificial ants that iteratively generate solutions by following probabilistic rules based on pheromone trails. These pheromone trails encode the collective information of the ants about the quality of the different solution components. The pheromone trails are dynamically modified according to the measured quality of the obtained solutions. The general scheme of an ACO algorithm is illustrated in Fig. 1. The metaheuristic begins with an initialization stage, where the pheromone trail is assigned a constant value, denoted as τ_0. At every step, the m virtual ants construct tours using the information from a problem-specific heuristic, η, and the pheromones, τ. Thus, at each iteration, m potential solutions (ant tours) are generated. Optionally, once the m ant tours are completed, the daemon actions procedure takes place. This procedure encompasses centralized actions unachievable for individual ants [8]. Subsequently, a rewarding strategy of the best tours is performed via the pheromone table update. This process is replicated until the defined stopping criterion is fulfilled and the algorithm yields the best tour found so far.

Different ACO algorithms may vary in their rules for updating the pheromones or their probabilistic transition rules. In this section, we begin by introducing the first ACO algorithm (ant system), due to its historical relevance and the fact that it provided inspiration for several extensions particularly tailored for improved performance. These extensions, considered the most successful ACO variants, are direct successors of AS and include AS_{Rank}[3], ACS [7], and MMAS [36].

Ant System (AS) was put forth in Marco Dorigo's doctoral dissertation [6] in the context of the TSP problem. As depicted in the general ACO scheme shown in Fig. 1, the AS algorithm can be divided into two main steps, namely, the solution construction of the ant tours and the pheromone updating. During the solution construction step, each of the m ant tours (candidate solutions) is incrementally constructed based on a choice rule combining the pheromone τ and heuristic η information:

$$p_{i,j}^{k} = \frac{\tau_{i,j}^{\alpha}\, \eta_{i,j}^{\beta}}{\sum_{l \in N_i^k} \tau_{i,j}^{\alpha}\, \eta_{i,j}^{\beta}} \quad , \quad j \in N_i^k \tag{1}$$

This probabilistic choice rule calculates the probability value that the ant k selects the next solution component from the feasible neighborhood N_i^k at the current solution component i. In the context of TSP, $p_{i,j}^k$ given by Eq. (1) represents the probability that ant k traverses the arc (i, j), going from city i to city j, with N_i^k denoting the list of cities still unvisited by the ants. Furthermore, α and β are control parameters accounting for the influence of the pheromone and the heuristic knowledge, respectively.

In ACO implementations for TSP, the pheromone trail $\tau_{i,j}$ influences the chance of traversing city j directly from i. The heuristic value associated with this transition $\eta_{i,j}$ is set inversely proportional to the computed distance between both cities.

Once each ant has finished its tour, the pheromone trail updating process (including evaporation and deposit mechanisms) takes place. First, all pheromone trails undergo a reduction by a constant factor following Eq. (2), where ρ is the pheromone evaporation rate, with $0 < \rho < 1$. This pheromone evaporation process is common to AS_{Rank} and MMAS.

$$\tau_{i,j} = (1-\rho)\tau_{i,j} \quad \forall (i, j) \tag{2}$$

Following this evaporation step, the ants deposit pheromones on edges traversed during their tour:

$$\tau_{i,j} = \tau_{i,j} + \sum_{k=1}^{m} \Delta\tau_{i,j}^k \quad \text{where} \quad \Delta\tau_{i,j}^k = \frac{1}{f(T^k)} \quad \forall (i, j) \in T^k \tag{3}$$

where $\Delta\tau_{i,j}^k$ denotes the pheromone quantity the k-th ant deposits on visited arcs. This amount is generally set inversely proportional to the fitness values of its solution, $f(T^k)$. In general, edges followed by many ants receive higher amounts, thus increasing the chance to be selected by ants at later iterations of the algorithm.

Rank-based Ant System (AS_{Rank}) is an AS extension introduced in [3] by considering the features of ranking and elitism into the pheromone update process. While in AS all ants lay down pheromone according to Eq. (3), in AS_{Rank} only w ants will do so: the best tour reached so far by the algorithm (T^{gb}), and the $(w-1)$ best ants at present iteration, as described by Eqs. (4) and (5), respectively.

$$\tau_{i,j} = \tau_{i,j} + w\,\Delta\,\tau_{i,j}^{gb} \quad \text{where} \quad \Delta\tau_{i,j}^{gb} = \frac{1}{f(T^{gb})} \quad \forall (i, j) \in T^{gb} \tag{4}$$

$$\tau_{i,j} = \tau_{i,j} + \sum_{r=1}^{w-1} (w-r)\,\Delta\,\tau_{i,j}^r \quad \text{where} \quad \Delta\tau_{i,j}^r = \frac{1}{f(T^r)} \quad \forall (i, j) \in T^r \tag{5}$$

The parameter w describes the number of ants depositing pheromones, while r accounts for the ant ranking according to the individual fitness (under the rule that the first rank $r = 1$ corresponds to the best fitness). According to Eq. (5), pheromone trails associated with the tours of the $(w-1)$ best ants at the present iteration are increased by an amount given by $(w - r)\,\Delta\tau^{r}_{i,j}$. Additionally, according to Eq. (4), the pheromone trails associated with the global-best solution T^{gb} found so far are reinforced by an amount given by $w\Delta\tau^{gb}_{i,j}$.

Ant Colony System (ACS), proposed in [7], is considered, along with MMAS, to be one of the best-performing extensions of AS [1, 4, 8, 9]. ACS incorporates modifications in the probabilistic transition rule and in the pheromone updating process compared to AS. On the one hand, ACS leverages the accumulated experience of the ants more strongly than AS does by using a stronger transition rule, given by Eq. (6):

$$j = \begin{cases} argmax_{l \in N^k_i}\{\tau_{il} \cdot \eta^{\beta}_{il}\} & q \leq q_0 \\ J & \text{otherwise} \end{cases} \tag{6}$$

where q is a uniform random variable, random variable J is taken according to the probability function given by Eq. (1) (with $\alpha = 1$), and q_0 is a parameter of ACS, with $0 < q_0 < 1$. With a probability of q_0, the next component j is assigned the best possible move according to the heuristic information and learned pheromone trails. Otherwise, j is determined through the transition rule of AS given by Eq. (1) with $\alpha = 1$.

On the other hand, for the pheromone updating process, ACS applies pheromone reinforcement and evaporation only to the edges belonging to the best-found ant tour:

$$\tau_{i,j} = (1 - \rho)\tau_{i,j} + \Delta\tau^{gb}_{i,j} \quad \text{where} \quad \Delta\tau^{gb}_{i,j} = \frac{1}{f(T^{gb})} \quad \forall (i, j) \in T^{gb} \tag{7}$$

where ρ is the pheromone evaporation parameter.

Besides the global pheromone trail updating rule of Eq. (7), ACS also applies a local updating rule during the construction loop, right after an ant crosses an edge (i, j):

$$\tau_{i,j} = (1 - \xi)\tau_{i,j} + \xi\tau_0 \tag{8}$$

where τ_0 denotes the initial value of the pheromone and ξ is a specific parameter of ACS, with $0 < \xi < 1$.

Max-Min Ant System (MMAS), reported in [36], implements two main modifications with respect to AS: a more aggressive pheromone reinforcement and the imposition of limits on the pheromone matrix. During pheromone reinforcement, given by Eq. (9), only the edges visited by the best ant are modified by an amount given by $\Delta\tau^{best}_{i,j}$, inversely proportional to the fitness. Depending on the implementation of the MMAS algorithm, it can consider either the iteration-best (denoted as T^{ib}) or the global-best (denoted as T^{gb}) ant tour.

$$\tau_{i,j} = \tau_{i,j} + \Delta\tau^{best}_{i,j} \text{ where } \Delta\tau^{best}_{i,j} = \frac{1}{f(T^{best})} \quad \forall (i, j) \in T^{best} \tag{9}$$

On the other hand, to prevent stagnation, MMAS considers minimum and maximum limits on pheromone trails (τ_{min} and τ_{max}). Therefore, after pheromone evaporation and reinforcement (Eqs. (2) and (9) resp.) at every iteration have taken place, minimum and maximum limits are imposed on all pheromone values, i.e., $\tau_{min} < \tau_{i,j} < \tau_{max}, \forall \tau_{i,j}$. The pheromone trail limits recommended by [8, 36] are $\tau_{max} = 1/(\rho f(T^{gb}))$ and $\tau_{min} = \tau_{max}(1 - \sqrt[n]{0.05})/(0.5n - 1)$, where n indicates the number of nodes (cities) in the graph. These pheromone bounds $[\tau_{min}, \tau_{max}]$ are used to prevent strict stagnation, in which all tours are identical. Instead, they might lead to a convergence situation where the edges of the best tour have pheromone value τ_{max}, while the remaining edges take the value τ_{min}. Under this situation, different ants can follow different tours, thereby avoiding stagnation. However, search space exploration becomes limited, thus hindering the chance of finding better solutions in next iterations. As a procedure to promote exploration, some implementations of MMAS consider occasional reinitializations of pheromone trails once the algorithm gets close to stagnation or whenever no enhanced tour is reached for a large number of consecutive iterative steps [8, 36].

Other ACO Algorithms and Hybridized Methods. There is an ongoing exploration into adapting ACO techniques for broader types of optimization problems. For instance, the so-called ACOR algorithm (standing for Ant Colony Optimization for Continuous Domains) [34] was introduced to extend the ACO paradigm to continuous optimization. Additionally, there is also a growing interest in dynamic problems, where certain aspects of the problem may vary during runtime. Among the newer ACO variants being developed to adapt to these changing scenarios, the population-based ACO (P-ACO) [14] proposes a mechanism where a list of solutions is preserved through iterations. Alongside the development of novel ACO techniques for various optimization problems, extensive research has been undertaken to enhance the ACO algorithm's performance for combinatorial problems through hybridization with other metaheuristic techniques. For instance, the Best-Worst Ant System (BWAS, [5]) considers both the best and worst solutions from an iteration into its pheromone update rule, and introduces a pheromone mutation strategy (a concept borrowed from evolutionary computation) along with a reinitialization of the pheromone trails as diversification strategies. Rokbani et al. [22] propose a hybridized metaheuristic for symmetric TSP, employing a simplified version of MMAS coupled with a 2-opt local search heuristic while optimizing ACO parameters using Particle Swarm Optimization (PSO). Another case of a hybridized metaheuristic is the extension of MMAS with 3-opt local search and heterogeneous ants proposed by [31], where each ant uses different values of heuristic influence and the pheromone parameters that can be modified by Gaussian mutation. In [28], a hybrid fuzzy ACO algorithm (FACO) is proposed for scheduling of virtual machines in high-performance cloud computing. For a general review of ACO's recent trends and state-of-the-art algorithms, see [9, 30].

3 Our Approach for Extending the Rank-Based Ant System

Regarding performance, all AS extensions reported above clearly show improvements with respect to the original AS [8]. The main advantage of the rank-based AS is its ability to rapidly discover reasonably good solutions during early iterations. However, its long-term results are generally inferior to those achieved by other AS extensions, such as MMAS and ACS [8]. For this reason, we propose an extension of AS_{Rank} that incorporates two diversification strategies aimed at enhancing the quality of the solutions found.

A distinctive characteristic of AS extensions is their assertive guidance of the ant's search, which heavily leverages the best solutions discovered so far. Unlike AS, where all ants lay down a pheromone quantity in proportion to the fitness, in the case of the ACS and MMAS methods, only a single ant (either the global best or the iteration best) deposits pheromones. Nevertheless, this intensified exploitation can lead to early convergence into a local minimum and to an undesirable situation known as stagnation. This state, with all ants following the same path, halts the exploration of new solutions, preventing the algorithm from escaping local minima. To counteract this, certain AS extensions such as MMAS or ACS include mechanisms to prevent search stagnation. ACS implements a local pheromone updating rule that applies evaporation exclusively on traversed paths, thereby promoting the exploration of unvisited arcs. Conversely, MMAS avoids search stagnation by imposing limits on the pheromone table. Additionally, some implementations of the MMAS procedure consider occasional pheromone reinitializations as a further means of increasing diversification.

AS_{Rank} also implements a more aggressive pheromone reinforcement than AS, where only w ants deposit pheromone, including the best-found solution and the $(w - 1)$ best ant tours of the present iteration. However, AS_{Rank} lacks any further strategies to prevent stagnation. In [25], we propose an extended version of AS_{Rank}, denoted hereafter as AS^{ps}_{Orank}, that incorporates two diversification strategies to prevent stagnation and enhance the quality of the obtained solutions. Namely, AS^{ps}_{Orank} incorporates a reinforcement strategy to promote original solutions and a new mechanism for pheromone smoothing, both described in the following subsections.

3.1 *Pheromone Reinforcement Rule Based on Originality Ranking*

The proposed ACO algorithm redefines the pheromone reinforcement criterion employed by AS_{Rank}, whose pheromone reinforcement rule, given by Eq. (5), assigns higher weights to arcs traversed by a selection of w ants based on their fitness ranking. In contrast, AS^{ps}_{Orank} introduces an additional dimension to the reinforcement criterion: the originality of an ant tour. A solution is considered more original if it explores edges in the graph that have been less frequently visited by the ants. By rewarding more original ant tours, the algorithm encourages ants to seek out less-traveled paths in the search for new, potentially superior solutions. In this subsection, we first define the fitness function for measuring the originality of the potential solutions and then explain how this criterion is incorporated into the pheromone update rule of AS^{ps}_{Orank}.

The distance between two ant tours can be expressed as the number of edges contained in just one of the tours, that is, the number of non-common edges [8]. This distance definition could be used for quantifying the originality of an ant tour compared with the others, by measuring the distance of all feasible pairs comprised by the ant tour and the others in the group. Unfortunately, such a measure shows to be computationally expensive [8].

The alternative originality function comes from the observation that the tours of ants traversing edges visited by fewer ants are "more original". Accordingly, the originality function f_o of tour T^k is computed as the inverse of the number of ants traversing each edge of T^k, as expressed by Eq. (10):

$$f_o(T^k) = \frac{1}{\sum\limits_{(i,j)\in T^k} \chi_{i,j}} \tag{10}$$

where $\chi_{i,j}$ denotes the number of ants traversing the edge connecting node i with node j. Therefore, whenever an ant tour T^k traverses edges chosen by a high number of ants (i.e., arcs with higher values of $\chi_{i,j}$), the originality value assigned by the originality fitness function $f_o(T^k)$ would be low.

This originality criterion is incorporated the pheromone reinforcement rule of AS^{ps}_{Orank} as follows. First, the $(w-1)$ best ants for the current iteration are selected according to their fitness. Next, their originality is evaluated according to Eq. (10), and an originality ranking is assigned ($r_o = 1, 2, ..., w-1$). Pheromone reinforcement of the $(w-1)$-best solution at this step is given by Eq. (11).

$$\tau_{i,j} = \sum_{r_o=1}^{w-1} (w - r_o) \bigtriangleup \tau^{r_o}_{i,j} \quad \text{where} \quad \Delta\tau^{r_o}_{i,j} = \frac{1}{f(T^{r_o})} \quad \forall (i, j) \in T^{r_o} \tag{11}$$

The pheromone evaporation and the reinforcement of the global-best tour of AS_{Rank} are maintained unmodified.

3.2 Pheromone Smoothing Strategy

A new strategy for pheromone smoothing is integrated into AS_{Rank} to help mitigate stagnation. To provide a clearer insight into this undesired situation, we present visual depictions of the pheromone trails obtained with AS_{Rank} during the optimization process of the TSPLIB instance *burma14* [29] after 0, 5, 10, and 100 iterations in Fig. 2. To this aim, we construct a pheromone table, actually a square matrix of

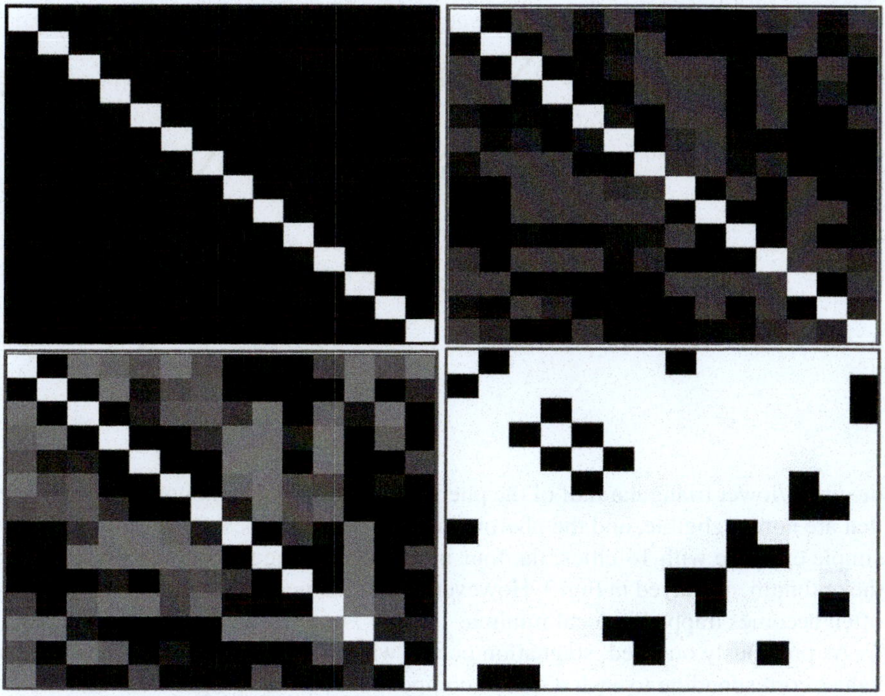

Fig. 2 (Top-bottom, left-right) Visualizations of the pheromone tables obtained by AS_{Rank} for the TSP instance *burma14* with fourteen cities are shown for 0, 5, 10, and 100 iterations, respectively

length n, with n indicating the number of nodes. For the specific instance *burma14*, the value of n is 14. Each element of the pheromone table, $\tau_{i,j}$, is related to the edge from node i to j. As *burma14* is a symmetric TSP instance, the pheromone table is symmetric, i.e., $\tau_{i,j} = \tau_{j,i}$. In Fig. 2, pheromone levels are illustrated using a gray-scale representation, where black (white) corresponds to the highest (lowest) pheromone value. At the initialization stage of AS_{Rank}, all pheromone values are set to a uniform value τ_0, excluding the diagonal ones (indicating no movement, $i = j$), which are not permitted (shown in white color in Fig. 2). After every iteration, all pheromones evaporate, as defined by Eq. (2), and reinforcement of some pheromone trails is carried out according to Eqs. (4) and (5). The effects of this pheromone updating process are observed in the representations of the pheromone trails at different iterations in Fig. 2. As the discrepancies among pheromone values increase, it becomes more likely for ants to select the edges chosen by higher quality ants in past iterations. By this reinforcement process, AS_{Rank} gains the ability to navigate promising regions of search space, leading to good solutions within a feasible timeframe. However, this process frequently results in early stagnation for AS_{Rank}, with all ants converging on the same path. This stagnation is evident after 100 iterations, see the bottom right side of Fig. 2. In this case, only 14 of the possible elements in

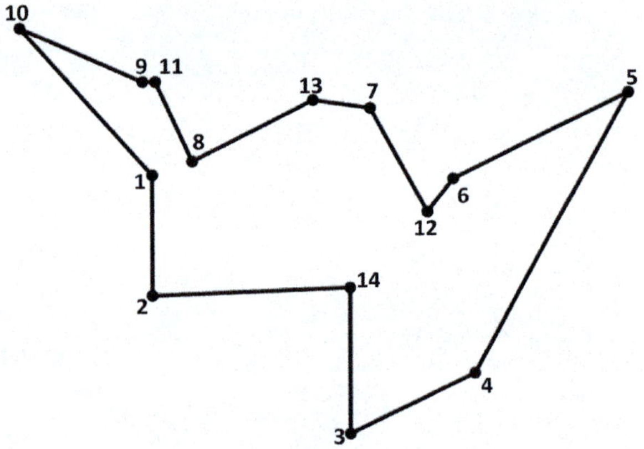

Fig. 3 Optimal tour for the *burma14* symmetric TSP instance with 14 cities

the upper/lower triangular half of the pheromone table retain amounts of pheromone that are non-negligible, and the pheromone table thus encodes a unique path. In this simple example with 14 cities, the tour in the pheromone table does correspond to the optimum, displayed in Fig. 3. However, this outcome is not the norm, and AS_{Rank} often becomes trapped in local minima.

As previously outlined, stagnation occurs when discrepancies in the pheromone values corresponding to each decision component become so significant that the ant tours, constructed in accordance with the probabilistic transition rule as specified by Eq. (1), end up being the same. This situation could be addressed by reassigning a constant value, such as τ_0, to the pheromone table, as proposed for MMAS [36]. However, when the pheromone table is reinitialized, all previously learned information is lost. As an alternative, the proposed pheromone smoothing mechanism re-scales the pheromone values to fit within the interval $[\gamma\tau_0, \tau_0]$, where $0 < \gamma < 1$, as expressed by Eq. (12). It is worth noting that if γ equals 1, pheromone smoothing becomes equivalent to pheromone reinitialization.

$$\tau = \gamma\tau_0 + \frac{\tau - min(\tau)}{max(\tau) - min(\tau)}(1-\gamma)\tau_0 \tag{12}$$

Integrating the pheromone smoothing strategy within the ACO framework requires to establish a criterion for its activation. For instance, in the context of MMAS, [36] employs occasional reinitialization when the λ-branching factor falls below a certain threshold. Despite serving as a robust indicator of the algorithm's exploration state, the λ-branching factor is computationally expensive [8], leading [36] to limit its computation to every 100 iterations. As an alternative, the standard deviation (denoted as σ_f) of the ant tours' fitness values is a more computationally efficient indicator of the diversity among the solutions. The σ_f of ant tours generated at each iteration reduces once the search process starts to exploit a specific region. And if σ_f reaches zero, it can be inferred that all ants are following an identical path (since it is rather improbable for different paths to have identical fitness), indicating stagnation. Because of its lower computational demand, σ_f is chosen as the triggering criterion for the pheromone smoothing mechanism in $\text{AS}^{\text{ps}}_{\text{Orank}}$.

4 Results

In this section, we analyze the performance of our extended variant of AS_{Rank}. The performance is assessed across two distinct combinatorial optimization problems and tested against state-of-the-art ACO metaheuristics, including AS_{Rank}, ACS, and MMAS. The section unfolds as follows: firstly, we describe the parameter settings utilized and provide an overview of our comparison methodology. Subsequently, we evaluate the ACO algorithms' performance on the classical optimization problem, TSP. Finally, we assess the performance of ACO algorithms for the MTS problem, a unique challenge that offers a valuable testing ground for assessing the algorithms' versatility.

4.1 Parameter Setup and Comparison Methodology

This subsection provides an overview of the parameter settings used in the experiments and describes the comparison methodology employed to assess the performance of ACO algorithms.

We selected the parameter settings based on recommendations from previous works [7, 8, 25, 36]. All algorithms were configured with: $\alpha = 1$ and $\beta = 2$, consistent with [7, 8, 25, 36]. For the rank-based ACO variants, we adopted specific parameters based on [8, 25]. We set $w = 6$ to determine the number of ants considered for the pheromone reinforcement rule, $\rho = 0.1$ for pheromone evaporation, and $\tau_0 = m/f(T^{gb})$ for the initialization of the pheromone trails. Additionally, for $\text{AS}^{\text{ps}}_{\text{Orank}}$, we considered $\gamma = 0.5$ for the pheromone smoothing mechanism, which was triggered whenever the standard deviation σ_f of the fitness of 80% of the solutions of the iteration reached zero, as described in [25]. For MMAS, the pheromone evaporation parameter was set to $\rho = 0.02$, and the iteration-best ant tour was selected for the pheromone reinforcement rule [8, 36]. We also implemented pheromone trails reinitialization to τ_{max} whenever an improved tour had not been found by MMAS for over 250 iterations, as outlined in [8]. Regarding the ant population size m, for TSP it was established to be the same as the node count for all the implemented algorithms except ACS, where we followed the recommendation of using 10 ants [7, 8]. For the MTS problem, we adopted a population size that scales with the UAV count (U) and the planning horizon (N), setting $m = 8 \cdot U \cdot N$, as suggested in [20, 24]. The stop criterion used was a predetermined maximum iteration count for MTS. For TSP, due to the lower number of ants in ACS, we opted for a maximum number of constructed tours to ensure fair comparison across all algorithms.

Accounting for the stochastic nature of ACO methods, each ACO algorithm was executed 40 times for each TSP or MTS instance. We analyzed the fitness values of the best solutions found from the 40 runs and presented the mean fitness values and standard deviations in comparative Tables 2 and 4. Furthermore, to assess statistical differences between the results, we conducted the Friedman test, a non-parametric alternative to ANOVA suitable for data with non-normal distributions. When statistically significant differences among the four ACO algorithms were observed, we conducted a post hoc Nemenyi test to identify specific pairwise differences between our algorithm and the other three algorithms. The Nemenyi test is a multiple comparison test that allows for the identification of significant differences between pairs of groups. The resulting p-values of these statistical tests are included in the last column of Tables 2 and 4.

4.2 Results over the TSP

ACO has a long-standing association with the TSP. The Ant system, being the pioneer ACO algorithm, was specifically tailored to tackle the TSP, and subsequent ACO algorithms have frequently been evaluated using TSP benchmarks. This subsection evaluates the performance of our ACO algorithm across several symmetrical TSP instances.

4.2.1 Traveling Salesman Problem

The Traveling Salesman Problem (TSP) endeavors to ascertain the minimal possible Hamiltonian loop in a given graph [23]. In this setting, each node corresponds to a city, and each edge correlates with the distance between any two cities. A TSP instance can be represented as an $n \times n$ distance matrix, wherein $d_{i,j}$ denotes the distance from city i to city j. The fitness of a closed tour T^k that traverses once each of the n cities is determined by the cumulative distances along the tour, as shown in Eq. (13).

$$fitness(T^k) = \sum_{\forall(i,j)\in T^k} d_{i,j} \qquad (13)$$

The problem is referred to as symmetric TSP if the directionality does not influence the distances between cities (i.e., $d_{i,j} = d_{j,i}, \forall(i, j)$). In contrast, if the distances are direction dependent, the problem is termed the asymmetric TSP.

4.2.2 ACO Formulation for TSP

When formulating ACO for the TSP, certain specific characteristics require special attention. As choosing nearby cities often yields favorable solutions for TSP, the inverse of the intercity distance is often adopted as an ad-hoc heuristic, i.e., $\eta_{i,j} = 1/d_{i,j}$. Additionally, within the TSP context, the neighborhood N_i^k includes the group of cities that remain unvisited. This approach enables ants to generate valid TSP tours by selecting unvisited cities at each iterative step.

Furthermore, when dealing with the symmetric TSP, the assumption is made that the pheromone matrix possesses symmetry, $\tau_{i,j} = \tau_{j,i}, \forall(i, j)$. Consequently, after applying the pheromone reinforcement rule that is specific to each ACO algorithm, symmetry is ensured by simply using Eq. (14).

$$\tau_{j,i} \leftarrow \tau_{i,j} \tag{14}$$

Furthermore, local search heuristics tailored for TSP such as the 2-opt heuristic [23] may also be included as part of the daemon actions to improve the results found by ACO algorithms. Nevertheless, as our goal is to analyze the impact of the proposed diversity strategies over TSP, we decided to implement all ACO algorithms without considering additional local search heuristics.

4.2.3 Comparative Results

Table 1 lists the main characteristics of the instances of TSP utilized for the experimental analysis: the names, the length of the optimal solution, and the number of cities/nodes. These instances are sourced from the benchmark of TSPLIB library [29].

Table 2 presents the comparative performance of MMAS, ACS, AS_{Rank}, and AS^{ps}_{Orank} over the TSP instances listed in Table 1. Each row of Table 2 provides metrics on the best solutions found by 40 runs of each ACO algorithm. These runs were conducted on each TSP instance with a stop criterion set to a maximum of 200,000 ant tours. Specifically, the third and fourth columns display the mean value (*Avg.*) and standard deviation (*Std.*) of the best solutions found during the 40 runs. For every instance, the best (lowest) mean fitness is highlighted in bold for easier comparison. The fifth column displays (*Best*) contain the length of the best ant tour obtained by each ACO algorithm across the 40 runs. The sixth column shows the percentage deviation of the average fitness (*Avg.*) from the length of the optimal solution, denoted as $p_{avg}(\%)$ and computed according to [11]. Analogously, the seventh column displays the percentage deviation of the fitness of the best solution among the 40 runs (*Best*) from the optimal one, denoted as $p_{best}(\%)$. Lastly, the eighth column displays the p-values resulting from the pairwise comparison of AS^{ps}_{Orank} with the other ACO algorithms, derived from the post hoc Nemenyi test, as described in Sect. 4.1.

In the case of state-of-the-art methods (MMAS, ACS, and AS_{Rank}), the results of Table 2 indicate that MMAS and ACS show similar good performance, often out-

Table 1 Symmetric traveling salesman problem instances from [29]

Instance	Optimal solution	Size (n)
brazil58	25,395	58
brg180	1,950	180
ch130	6,110	130
ch150	6,528	150
eil101	629	101
gr48	5,046	48
gr202	40,160	202
kroA100	21,282	100
kroB200	29,437	200
pr76	108,159	76
pr264	49,135	264
st70	675	70
tsp225	3,916	225

performing AS_{Rank}. MMAS outperforms ACS in certain instances (such as *brg180*, *ch130*, *ch150*, and *kroA100*), whereas in others (*eil101*, *gr48*, *gr202*, or *pr264*) ACS exhibits superior results. In most cases, AS_{Rank} produces solutions inferior to both MMAS and ACS. These observations are consistent with the conclusions drawn from similar comparative studies [8].

Furthermore, the results from the experiments presented in Table 2 highlight the superior performance of AS^{ps}_{Orank}, which achieves lower mean fitness than other ACO methods in 11 out of 13 instances. Specifically, the improvement over MMAS is statistically significant in three instances (*brazil58*, *gr48*, and *pr264*), over ACS in six instances (*brg180*, *ch130*, *ch150*, *kroA100*, *pr76*, and *st70*), and over both algorithms in *gr202*, *kroB200*, and *tsp225*. From the comparison of the proposed algorithm with AS_{Rank}, we can conclude that the proposed diversification strategies clearly benefit the algorithm's performance. Specifically, AS^{ps}_{Orank} consistently achieves significantly shorter tours than AS_{Rank} across all the TSP instances.

4.3 Results over the Minimum Time Search Problem

In this section, we assess the performance of our ACO algorithm in comparison with other ACO variants over several MTS scenarios, where the objective is to optimize the trajectories of several UAVs in order to determine the location of an undisclosed target.

Table 2 Comparative results of ACO algorithms for TSP instances

Instance	Alg.	Avg.	Std.	Best	$p_{avg}(\%)$	$p_{best}(\%)$	p-value
brazil58	MMAS	25,634	45	25,395	0.94	0.00	0.001
	ACS	25,453	101	25,395	0.23	0.00	0.799
	AS_{Rank}	25,632	137	25,400	0.93	0.02	0.001
	AS^{ps}_{Orank}	**25,396**	2	25,395	0.00	0.00	–
brg180	MMAS	**1,951**	3	1,950	0.05	0.00	0.900
	ACS	1,958	7	1,950	0.38	0.00	0.009
	AS_{Rank}	1,957	7	1,950	0.36	0.00	0.041
	AS^{ps}_{Orank}	1,952	4	1,950	0.10	0.00	–
ch130	MMAS	6,197	25	6,132	1.43	0.36	0.329
	ACS	6,217	50	6,139	1.75	0.47	0.003
	AS_{Rank}	6,249	41	6,185	2.28	1.23	0.001
	AS^{ps}_{Orank}	**6,182**	36	6,128	1.18	0.29	–
ch150	MMAS	6,574	15	6,548	0.70	0.31	0.190
	ACS	6,610	46	6,543	1.25	0.23	0.001
	AS_{Rank}	6,620	33	6,550	1.42	0.34	0.001
	AS^{ps}_{Orank}	**6,563**	13	6,543	0.53	0.23	–
eil101	MMAS	644	11	633	2.31	0.64	0.175
	ACS	639	5	630	1.57	0.16	0.900
	AS_{Rank}	644	5	632	2.33	0.48	0.001
	AS^{ps}_{Orank}	**637**	5	629	1.34	0.00	–
gr48	MMAS	5,101	32	5,063	1.09	0.34	0.013
	ACS	5,087	23	5,046	0.81	0.00	0.429
	AS_{Rank}	5,136	47	5,072	1.79	0.52	0.001
	AS^{ps}_{Orank}	**5,076**	17	5,049	0.60	0.06	–
gr202	MMAS	42,699	483	41,778	6.32	4.03	0.001
	ACS	41,689	445	40,985	3.81	2.05	0.001
	AS_{Rank}	41,810	394	41,120	4.11	2.39	0.001
	AS^{ps}_{Orank}	**41,108**	277	40,539	2.36	0.94	–
kroA100	MMAS	21,376	87	21,282	0.44	0.00	0.109
	ACS	21,492	256	21,282	0.99	0.00	0.001
	AS_{Rank}	21,715	258	21,320	2.03	0.18	0.001
	AS^{ps}_{Orank}	**21,347**	106	21,282	0.31	0.00	–
kroB200	MMAS	30,401	327	29,922	3.28	1.65	0.001
	ACS	30,231	368	29,554	2.70	0.40	0.001
	AS_{Rank}	30,486	232	29,941	3.56	1.71	0.001
	AS^{ps}_{Orank}	**29,870**	211	29,540	1.47	0.35	–

(continued)

Table 2 (continued)

Instance	Alg.	Avg.	Std.	Best	$p_{avg}(\%)$	$p_{best}(\%)$	p-value
pr76	MMAS	109,807	1150	108,159	1.52	0.00	0.900
	ACS	**108,974**	951	108,159	0.75	0.00	0.065
	AS_{Rank}	111,805	981	110,407	3.37	2.08	0.001
	AS^{ps}_{Orank}	109,594	859	108,159	1.33	0.00	–
pr264	MMAS	50,456	749	49,491	2.69	0.72	0.001
	ACS	49,831	589	49,144	1.42	0.02	0.750
	AS_{Rank}	50,999	1044	49,377	3.79	0.49	0.001
	AS^{ps}_{Orank}	**49,661**	655	49,144	1.07	0.02	–
st70	MMAS	683	7	675	1.19	0.00	0.121
	ACS	684	7	675	1.33	0.00	0.032
	AS_{Rank}	689	8	679	2.14	0.59	0.001
	AS^{ps}_{Orank}	**680**	5	675	0.79	0.00	–
tsp225	MMAS	4,066	39	3,999	3.84	2.12	0.001
	ACS	4,023	41	3,940	2.73	0.61	0.001
	AS_{Rank}	4,043	34	3,978	3.25	1.58	0.001
	AS^{ps}_{Orank}	**3,954**	20	3,916	0.98	0.00	–

4.3.1 The Minimum Time Search Problem

Recent years have witnessed a marked upsurge in research interest in mission planning for Unmanned Aerial Vehicles (UAVs) due to their numerous advantages across various applications. One such application is the MTS problem, which involves determining the optimal trajectories for one or more UAVs tasked with finding a target with an unknown location in the shortest possible time. The MTS problem falls within the broader category of problems known as Probabilistic Search. In these problems, a probabilistic approach is employed to handle inherent uncertainties. Specifically, probability models are used to represent the uncertain knowledge about the target's locations, dynamics, and sensor measurements.

The specific location of the target is still unknown and is modeled through a random variable ν^t. In this search problem, it is assumed that a probability function, representing the accessible knowledge about the target's whereabouts, is obtainable prior to the initiation of the search. Such a function can be depicted through a grid-based probability map, also referred to as a "belief" and represented by the notation $b(\nu^0)$. The searching area is segmented into a grid G_Ω composed of square cells of dimension $w_x \times w_y$. Within the belief map, each cell has an associated probability reflecting the possible presence of the target within that cell. Additionally, whenever the target is not static, its dynamic information can be modeled using a target motion model $P(\nu^t|\nu^{t-1})$, denoting the probability of the target moving from a cell ν^{t-1} to other location ν^t. Furthermore, the sensor model allows for the management of uncertainty inherent in sensor measurements. The probabilistic model $P(z_u^t = D|s_u^t, \nu^t)$

represents the probability of target detection ($z_u^t = D$) knowing the locations of the target ν^t and the position of the u-th UAV s_u^t. A common assumption in probabilistic search is the consideration of two kinds of sensor measurements: target detection ($z_u^t = D$) and its opposite ($z_u^t = \overline{D}$). Then, the probability of non-target detection holds $P(z_u^t = \overline{D}|s_u^t, \nu^t) = 1 - P(z_u^t = D|s_u^t, \nu^t)$. To evaluate our approach we use an ideal sensor model with perfect detection probability equal to one if the target and the UAV are over the same cell, and 0 otherwise, as in [20, 24, 26].

We discretize the maneuverability of the UAVs in terms of time and space, enabling each UAV to make decisions and take measurements at distinct points in time called time steps. These time steps are spaced evenly by ΔT, representing the duration required for a UAV to fly from one cell to an adjacent one and take a measurement. The U UAVs maintain a constant flying height, and their locations are restricted to the centers of the cells in the grid G_Ω. The dynamic model we consider allows each UAV to move from its current cell s_t to an adjacent cell corresponding to the eight cardinal points labeled as N, NE, E, SE, S, SW, W, and NW. Consequently, the search paths are comprised of sequences of neighboring cells at a constant altitude, suitable for rotary-wing UAVs to follow. Taking into account the limited duration of the search mission, we determine the temporal horizon N, representing how many time steps are allowed for the search mission, with step size ΔT. Because of the determinism present in the UAVs' dynamic model, the search trajectories can be defined either by the list of traversed cells $s_{1:U}^{0:N}$ or by the list of motion actions $c_{1:U}^{1:N}$ along with the starting positions of the UAVs $s_{1:U}^0$. We remark that the UAVs are not allowed to move outside the search zone.

It is worth noting that the number of candidate search trajectories grows exponentially with N and U (the number of drones), resulting in $8^{N \cdot U}$ potential solutions, leading to an NP-hard problem [37]. We also seek to obtain high-quality UAV trajectories within a reasonable timeframe. Therefore, the problem is often simplified and addressed using generic approximate optimization methods such as cross-entropy optimization [20], or metaheuristics such as ACO [24, 26].

The primary objective of MTS is to minimize the time required to locate the target. Consequently, the expected time value for target detection, denoted as $ET(s_{1:U}^{0:t})$, is commonly selected as a fitness criterion. $ET(s_{1:U}^{0:t})$ can be obtained using Eqs.(15) and (16), as in [20, 24, 26].

$$ET(s_{1:U}^{0:\infty}) = \sum_{t=0}^{\infty} \sum_{\nu^t \in G_\Omega} \tilde{b}(\nu^t) \tag{15}$$

$$\tilde{b}(\nu^t) = \prod_{u=1}^{U} P(z_u^t = \overline{D}|\nu^t, s_u^t) \sum_{\nu^{t-1} \in G_\Omega} P(\nu^t|\nu^{t-1})\tilde{b}(\nu^{t-1}) \tag{16}$$

The expected time (ET) is obtained by summing up all the cell values of $\tilde{b}(\nu^t)$ for every time step from zero to infinity, as described by Eq. (15). The updated unnormalized belief $\tilde{b}(\nu^t)$ is obtained through Eq. (16) by recursively updating the previous time step $\tilde{b}(\nu^{t-1})$, using the target's dynamical information modeled by $P(\nu^t|\nu^{t-1})$ and the non-detection likelihood information represented by $P(z_u^t = \overline{D}|\nu^t, s_u^t)$. For the starting case ($t = 0$), $\tilde{b}(\nu^0)$ is set equal to the initial probability map $b(\nu^0)$. In this way, Eqs. (15) and (16) enable the computation of the ET in terms of the input models: the initial probability map $b(\nu^0)$, the target dynamical model $P(\nu^t|\nu^{t-1})$, and the sensor model $P(z_u^t|\nu^t, s_u^t)$. It should be noted that to transform the expected time from time steps to actual time units, the ET step should be scaled by ΔT for each two subsequent sensor samples.

Equation (17) estimates the ET of a search trajectory by summing all the cell values of $\tilde{b}(\nu^t)$ at each time step up to a fixed value N. This truncated expression of the expected time is often used as the fitness function in probabilistic search problems [20, 24]. We refer to it simply as expected time or ET.

$$fitness(s_{1:U}^{0:N}) = \sum_{t=0}^{N} \sum_{\nu^t \in G_\Omega} \tilde{b}(\nu^t) \tag{17}$$

In this context, $fitness(s_{1:U}^{0:N})$ denotes the fitness value of a search trajectory $s_{1:U}^{0:N}$.

4.3.2 ACO Formulation for MTS

As described in the previous subsection, MTS is a combinatorial optimization problem seeking to determine the best possible search trajectories for U UAVs. This subsection outlines the essential aspects needed to approach the MTS problem using ACO metaheuristics, specifically covering the definition of a suitable solution codification, the encoding of learned information through pheromones, and the adoption of a constructive MTS heuristic.

The search space in MTS can be depicted as a graph, wherein each node matches a cell in the grid-like search area G_Ω, and the edges connect adjacent cells following the eight cardinal points. It is important to note that, unlike TSP, where all nodes are fully connected, in MTS, each node maintains connectivity solely with its eight adjacent nodes (cells). Additionally, while TSP tours form closed loops and the initial position of each ant's tour is irrelevant, MTS requires generating search trajectories that originate from the UAVs' initial positions. Given the starting positions of the U UAVs, the candidate solutions are represented by the tours of each ant. These tours are determined by the cardinal control commands $c_u^t = \{1, 2, 3, 4, 5, 6, 7, 8\}$ taken by the U UAVs at each time step up to N. The discrete codification of the candidate solutions can be codified as

$$\overbrace{c_1^1, c_1^2, ..., c_1^N}^{\text{drone 1}}, \overbrace{c_2^1, c_2^2, ..., c_2^N}^{\text{drone 2}}, ..., \overbrace{c_U^1, c_U^2, ..., c_U^N}^{\text{drone U}} \tag{18}$$

Considering the cardinal dynamic model that enables movements between adjacent cells in G_Ω, these candidate solutions $c_{1:U}^{1:N}$ can be easily converted into a sequence of visited cells $s_{1:U}^{0:N}$ that correspond to an ant tour T^k. This codification offers the advantage that the majority of the combinations constitute feasible search trajectories, with the exception of sequences that would cause the UAVs to exit the search area.

Following the approach used in [24], the pheromone table τ gains knowledge of the best control actions for each UAV at every time step. Consequently, the pheromone table is a 3D matrix with dimensions $(8, N, U)$, with its elements $\tau[a, t, u]$ arranged in rows corresponding to the control actions (a), columns for time steps (t), and depth for the UAVs (u). This codification allows for the storage of knowledge acquired for each UAV independently, which is advantageous as UAVs in MTS often distribute themselves and follow distinct paths. The pheromone evaporation and reinforcement rules from the ACO algorithms presented in Sect. 2 are adapted to the specific pheromone encoding used in MTS. For instance, the pheromone evaporation rule, originally defined by Eq. (2), is modified to Eq. (19), which states that all the elements in the pheromone table are evaporated according to the parameter ρ.

$$\tau[c, t, u] \leftarrow (1 - \rho) \cdot \tau[c, t, u] \quad \forall (c, t, u) \tag{19}$$

As another example, the pheromone reinforcement rule of MMAS, originally defined by Eq. (9), is now defined by Eq. (20). In this adapted rule, only the actions $^*c_u^t$ taken at time step t by the u-th UAV corresponding to the best ant tour T^{best} at current iteration are reinforced. The reinforcement is performed by adding a quantity $\Delta\tau_{i,j}^{best}$ inversely proportional to the best ant tour's fitness, which is the expected time obtained by Eq. (17).

$$\tau[^*c_u^t, t, u] = \tau[^*c_u^t, t, u] + \Delta\tau_{i,j}^{best}, \text{ where } \Delta\tau_{i,j}^{best} = \frac{1}{ET(T^{best})} \; \forall ^*c_u^t \in T^{best} \tag{20}$$

The constructive heuristic, which is specifically tailored for MTS, that we adopted is the one used in [24, 26]. This MTS heuristic, as given by Eq. (21), computes the heuristic value given to a control action a, based on actual cell s_u^t of the u$-th$ UAV at time step t. It aggregates the values of the updated belief $\tilde{b}(\nu^t)$, weighted by a function of the distance between s_u^t and the still reachable cells j within the triangle corresponding to the cardinal control action a, i.e., $j \in triangle(a, s_u^t, N - t)$. Figure 4 illustrates the cells taken into account to compute $\eta(a, s_u^t, t)$, which are the cells contained within $triangle(a, s_u^t, N - t)$. The weighting function $f(distance(i, j))$ assigns greater weights to cells j of $\tilde{b}(\nu^t)$ nearer to the actual position of the drone s_u^t. Through this procedure, the MTS heuristic enhances the chances of choosing cardinal control commands that guide the drones toward closer regions with a higher probability of target presence, within the remaining $(N - t)$ time steps.

$$\eta(a, s_u^t, t) = \sum_{j \in triangle(a, s_u^t, N-t)} f(distance(i, j)) \, \tilde{b}(\nu^t = j) \tag{21}$$

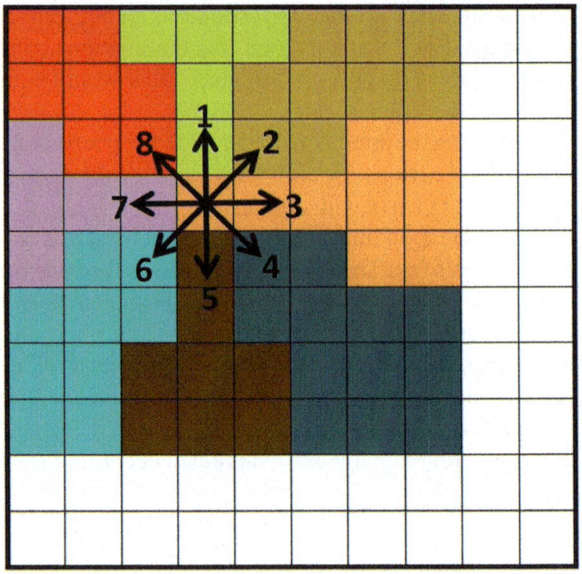

Fig. 4 The cells j within each $triangle(a, s_u^t, N-t)$ employed to assign heuristic values to each action a for a drone located at s_u^t are shown with different colors. The drone is located at the center of the black arrows

During the solution construction loop, the control actions of each UAV are sequentially obtained based on the probabilistic rule specified by Eq. (22).

$$P(a,t,u) = \frac{[\tau[a,t,u]]^{\alpha}\left[\eta(a,s_u^t,t)\right]^{\beta}}{\sum\limits_{a=1:8}[\tau[a,t,u]]^{\alpha}\left[\eta(a,s_u^t,t)\right]^{\beta}} \tag{22}$$

Equation (22) is an adaptation of the transition rule initially given by Eq. (1) that determines the probability that the u-th UAV takes the cardinal action a at time step t. It is important to note that in MTS, the feasible neighborhood consists of the eight cardinal actions, excluding the cells located at the border where $P(a,t,u)$ is set to zero to confine the trajectories within the search area.

For additional information on the formulation of MTS with ACO, we suggest consulting [24]. This work is noteworthy because it represents the first study to employ ACO algorithms, specifically MMAS, to address the MTS problem.

4.3.3 Comparison Results

For the analysis of the ACO algorithms over the MTS problem, we selected the benchmark used in [24]. This benchmark consists of six search scenarios that vary in the number of UAVs (U), the initial beliefs, the target motion models, and the planning horizon (N). These properties are summarized in Table 3 and depicted in Fig. 5, where the initial UAVs' locations are marked by gray circles, the initial beliefs are displayed

Table 3 Main characteristics of search scenarios from MTS benchmark from [24]

Instance	UAVs (U)	Dynamic target	Planning horizon (N)
Scenario A	1	No	10
Scenario B	1	Yes	20
Scenario C	2	Yes	10
Scenario D	3	Yes	20
Scenario E	2	Yes	10
Scenario F	1	No	20

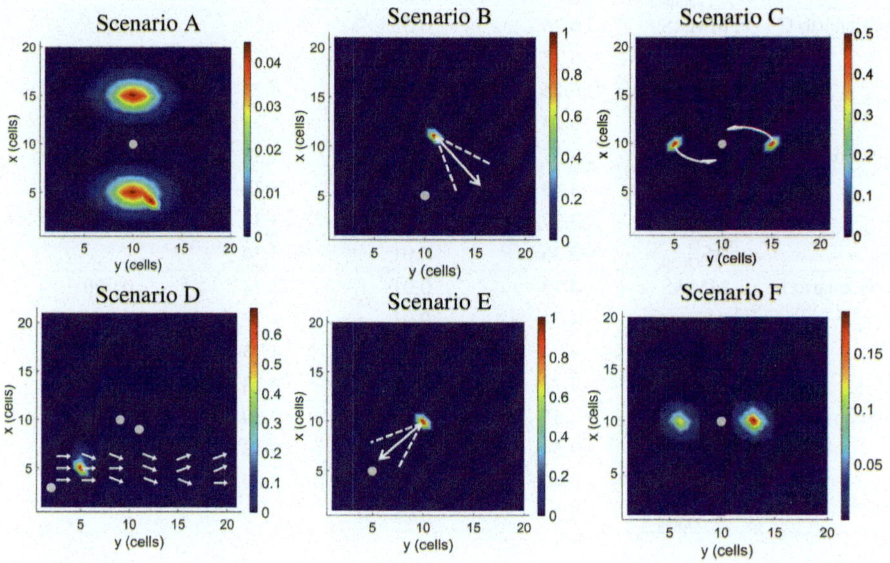

Fig. 5 MTS search scenarios

with colored maps, and the target models are marked with white arrows. For instance, in Scenario A the planning horizon is equal to 10, and the initial target belief has two high-probability areas equidistant from the initial position of the unique drone. The challenge of this scenario arises from the minor discrepancy in fitness between the trajectories that lead to the probability zones, making the likelihood of the area located in the south only marginally higher. In Scenario C, two UAVs starting from the same initial position carry out the search for a dynamic target. The complexity of this scenario lies in the circular dynamics of two initially clustered high-probability regions. Scenario D models complex target dynamics that simulate the dynamics of a lost boat in the ocean, derived from a stochastic wind map. In this scenario, the two UAVs starting from central positions must intercept the belief, while the UAV initially located at the bottom-left corner tracks the target's movements. The results obtained by the ACO algorithms after 300 iterations across the search scenarios are shown in

Table 4 Comparative results of ACO algorithms for MTS scenarios from [24]

Instance	Algorithm	Avg.	Std.	Best	p-value
Scenario A	MMAS	**8.96**	0.01	8.95	0.900
	ACS	**8.96**	0.01	8.95	0.653
	AS_{Rank}	8.97	0.02	8.95	0.019
	AS^{ps}_{Orank}	**8.96**	0.02	8.95	–
Scenario B	MMAS	9.56	0.03	9.47	0.455
	ACS	9.82	0.25	9.56	0.001
	AS_{Rank}	9.56	0.04	9.55	0.677
	AS^{ps}_{Orank}	**9.55**	0.02	9.55	–
Scenario C	MMAS	6.25	0.02	6.23	0.036
	ACS	6.27	0.03	6.23	0.001
	AS_{Rank}	6.25	0.02	6.23	0.001
	AS^{ps}_{Orank}	**6.24**	0.01	6.23	–
Scenario D	MMAS	3.37	0.02	3.36	0.604
	ACS	3.43	0.03	3.36	0.001
	AS_{Rank}	3.37	0.01	3.36	0.604
	AS^{ps}_{Orank}	**3.36**	0.01	3.36	–
Scenario E	MMAS	**4.73**	0.01	4.73	0.900
	ACS	4.74	0.01	4.73	0.005
	AS_{Rank}	4.74	0.01	4.73	0.099
	AS^{ps}_{Orank}	**4.73**	0.01	4.73	–
Scenario F	MMAS	9.53	0.19	9.09	0.004
	ACS	9.54	0.25	9.18	0.006
	AS_{Rank}	9.47	0.25	9.09	0.244
	AS^{ps}_{Orank}	**9.34**	0.25	9.09	–

Table 4. For each of the six search scenarios, the table displays the average expected time, calculated using Eq. (17), along with its standard deviation. For each search scenario, the best (lowest) mean fitness is highlighted in bold for easier comparison. The table also includes best-found solution fitness by the 40 independent runs and the p-values from the pairwise comparison of each ACO algorithm with AS^{ps}_{Orank}, derived from the post hoc Nemenyi test as described in Sect. 4.1.

The results presented in Table 4 remark the superior performance of AS^{ps}_{Orank} within this MTS benchmark. Following AS^{ps}_{Orank} in terms of performance are MMAS, AS_{Rank}, and ACS. Specifically, AS^{ps}_{Orank} consistently achieves lower or at least equal mean fitness compared to all other ACO algorithms across all scenarios. Notably, AS^{ps}_{Orank} obtains statistically significant improvements over MMAS in Scenarios C and F, and over ACS in search Scenarios B, C, D, E, and F.

Furthermore, the proposed ACO algorithm obtains lower mean fitness than the original AS_{Rank} across all six scenarios, exhibiting statistically significant improvements in Scenarios A, C, and E. These results indicate that the innovative strategies (originality reinforcement, pheromone smoothing) notably enhance the performance of AS_{Rank} in MTS.

5 Conclusions

In this chapter, we have provided a thorough discussion and analysis of our extended rank-based ant system, which integrates two novel strategies for diversification: (1) originality reinforcement; (2) pheromone smoothing. The originality reinforcement strategy aims to reward innovative solutions by promoting exploration of new areas within the search space, while the pheromone smoothing strategy helps avoid stagnation, enabling the algorithm to escape from local optimal solutions.

In addition to situating our proposed algorithm within the context of existing ACO algorithms and providing a detailed description of it, we rigorously evaluated the approach on two diverse combinatorial optimization problems: the classic TSP and the more complex MTS problem. Each problem required a tailored formulation of the ACO algorithm to address their unique characteristics and constraints.

The experimental results over both combinatorial optimization problems clearly demonstrate the improvement in performance compared to the rank-based ant system, highlighting the effectiveness of the two proposed diversification strategies. Furthermore, our proposed ACO algorithm consistently achieves better or at least similar performance compared to widely used ACO algorithms ACS and MMAS.

As future research lines, we are considering the analysis of the proposed approach's performance on larger TSP instances. We will explore specific modifications commonly used to deal with large TSP instances, such as candidate lists or fixed-radius nearest neighbor search [8]. Additionally, we intend to expand our approach to dynamic optimization problems, such as dynamic TSP [14], where adaptability and efficiency are crucial in addressing changing scenarios.

Acknowledgements The authors would like to thank the financial support from the European Union's Horizon 2020 research and innovation programme, Marie Sklodowska-Curie action, RISE program, of the project PDE-GIR with grant agreement reference number 778035, and also from the Agencia Estatal de Investigación (AEI) of the Spanish Ministry of Science and Innovation, for the grant of the Computer Science National Program with reference number #PID2021-127073OB-I00 of the MCIN/AEI/10.13039/501100011033/FEDER, EU.

References

1. Blum C (2005) Ant colony optimization: introduction and recent trends. Phys Life Rev 2(4):353–373
2. Bonabeau E, Dorigo M, Theraulaz G (1999) Swarm intelligence: from natural to artificial systems. Oxford University Press

3. Bullnheimer B, Hartl RF, Strauss C (1999) A new rank-based version of the ant system: a computational study. CentL Eur J Oper Res Econ 7(1):25–38
4. Cordón O, Herrera F, Stützle T (2002) A review on the ant colony optimization metaheuristic: basis, models and new trends. Mathw & Soft Comput 9(2–3)
5. Cordón O, Viana IFD, Herrera F (2002) Analysis of the best-worst ant system and its variants on the QAP. In: International workshop on ant algorithms. Springer, Berlin, Heidelberg, pp. 228–234
6. Dorigo M (1992) Optimization, learning and natural algorithms. PhD thesis, Politecnico di Milano, Italy
7. Dorigo M, Gambardella LM (1997) Ant colony system: a cooperative learning approach to the traveling salesman problem. IEEE Trans Evol Comput 1(1):53–66
8. Dorigo M, Stützle T (2004) Ant colony optimization. MIT Press, Cambridge, Massachusetts, USA
9. Dorigo M, Stützle T (2019) Ant Colony Optimization: Overview and recent advances. In: Handbook of metaheuristics. Springer International Publishing, pp 311–351
10. Engelbrecht A (2005) Fundamentals of computational swarm intelligence. Wiley & Sons
11. Gao W (2020) New ant colony optimization algorithm for the traveling salesman problem. Int J Comput Intell Syst 13(1):44–55
12. Glover F (1989) Tabu search—Part I. INFORMS J Comput 1(3):190–206
13. Glover F (1990) Tabu search—Part II. INFORMS J Comput 2(1):4–32
14. Guntsch M, Middendorf M (2002) Applying population based ACO to dynamic optimization problems. In: Ant algorithms: third international workshop, ANTS (2002) Brussels, Belgium, September 12–14, 2002 proceedings 3. Springer, Berlin, Heidelberg, pp 111–122
15. Goldberg DE (1989) Genetic algorithms in search. Optimization and machine learning. Addison-Wesley Longman Publishing Co., Inc, Boston, MA, USA
16. Holland JH (1975) Adaptation in natural and artificial systems. University of Michigan Press, Ann Arbor
17. Kennedy J, Eberhart RC (1995) Particle swarm optimization. In: Proceedings of IEEE international conference on neural networks, vol. 4, pp 1942–1948
18. Kennedy J, Eberhart RC (2001) Swarm intelligence. Morgan Kaufmann
19. Kirkpatrick S, Gelatt CD Jr, Vecchi MP (1983) Optimization by simulated annealing. Sci 220(4598):671–680
20. Lanillos P, Besada-Portas E, Pajares G, Ruz JJ (2012) Minimum time search for lost targets using cross entropy optimization. In: IEEE/RSJ international conference on intelligent robots and systems. IEEE, pp 602–609
21. Maniezzo V, Dorigo M, Colorni A (1996) The ant system: optimization by a colony of cooperating agents. IEEE Trans Syst Man Cybern Part B 26(1):29–41
22. Rokbani N, Kromer P, Twir I, Alimi AM (2019) A new hybrid gravitational particle swarm optimisation-ACO with local search mechanism, PSOGSA-ACO-Ls for TSP. Int J Intell Eng Inform 7(4):384–398
23. Ouaarab A (2020) Review of tour generation for solving traveling salesman problems. In: Yang XS, Zhao YX (eds) Nature-inspired computation in navigation and routing problems. Springer tracts in nature-inspired computing. Springer, Singapore
24. Pérez-Carabaza S, Besada-Portas E, Lopez-Orozco JA, de la Cruz JM (2018) Ant colony optimization for multi-UAV minimum time search in uncertain domains. Appl Soft Comput 62:789–806
25. Pérez-Carabaza S, Gálvez A, Iglesias A (2022) Rank-based ant system with originality reinforcement and pheromone smoothing. Appl Sci 12(21):11219
26. Pérez-Carabaza S, Scherer J, Rinner B, López-Orozco JA, Besada-Portas E (2019) UAV trajectory optimization for Minimum Time Search with communication constraints and collision avoidance. Eng Appl Artif Intell 85:357–371
27. Price K, Storn RM, Lampinen JA (2005) Differential evolution: a practical approach to global optimization. Springer

28. Ragmani A, Elomri A, Abghour N et al (2020) FACO: a hybrid fuzzy ant colony optimization algorithm for virtual machine scheduling in high-performance cloud computing. J Ambient Intell HumIzed Comput 11:3975–3987. https://doi.org/10.1007/s12652-019-01631-5
29. Reinelt G (1991) TSPLIB-A traveling salesman problem library. ORSA J Comput 3(4):376–384
30. Rokbani N, Kumar R, Abraham A, Alimi AM, Long HV, Priyadarshini I, Son LH (2021) Bi-heuristic ant colony optimization-based approaches for traveling salesman problem. Soft Comput 25:3775–3794
31. Tuani AF, Keedwell E, Collett M (2020) Heterogenous adaptive ant colony optimization with 3-opt local search for the travelling salesman problem. Appl Soft Comput 97:106720
32. Shi Y, Eberhart RC (1998) A modified particle swarm optimizer. In: Proceedings of IEEE international conference on evolutionary computation, IEEE-CEC, pp 69–73
33. Siarry P (ed) (2016) Metaheuristics, vol 71. Springer, Berlin/Heidelberg, Germany
34. Socha K, Dorigo M (2008) Ant colony optimization for continuous domains. Eur J Oper Res 185(3):1155–1173
35. Storn R, Price K (1997) Differential evolution—a simple and efficient heuristic for global optimization over continuous spaces. J Glob Optim 11(4):341–359
36. Stützle T, Hoos HH (2000) MAX-MIN ant system. Futur Gener Comput Syst 16(8):889–914
37. Trummel KE, Weisinger JR (1986) The complexity of the optimal searcher path problem. Oper Res 34(2):324–327

Chapter 3
Exploring Ant Colony Optimization for Feature Selection: A Comprehensive Review

A. Hashemi and M. B. Dowlatshahi

1 Introduction

Today, large-scale data are produced at high speed in various real-world applications, including scientific research, business, and social networking. Much of these data cannot be processed by typical database programs. This type of data, generated in large volumes and dimensions, is referred to as big data. The reason for the large amount of data in many of these datasets is the presence of many features that pose challenges to machine learning algorithms. This problem occurs especially in classification tasks for big data. On the other hand, not all features in these datasets are useful [1, 2].

Not all features in the dataset are suitable for the training phase, and many are redundant or irrelevant. The presence of unrelated features misleads the learning algorithm and reduces the classification accuracy. On the other hand, some features are similar to other features and do not contain new information. Without performing preprocessing on the datasets and allowing the existence of unrelated and redundant features, in addition to reducing the accuracy of the model, it causes problems such as long learning time and higher computational complexity to the model. As a well-known preprocessing technique, feature selection minimizes the impact of these features. Feature selection selects the most relevant and unique features with minimal similarity to other features to improve the predictive power of the model and reduce computational complexity and execution time [3–5].

Considering the extensive use of feature selection in a variety of engineering problems, it is important to provide effective methods in this area. If we consider d

A. Hashemi · M. B. Dowlatshahi (✉)
Faculty of Engineering, Department of Computer Engineering, Lorestan University, Khorramabad, Iran
e-mail: dowlatshahi.mb@lu.ac.ir

A. Hashemi
e-mail: hashemi.am@fe.lu.ac.ir

N. Dey (ed.), *Applications of Ant Colony Optimization and its Variants*, Springer Tracts in Nature-Inspired Computing, https://doi.org/10.1007/978-981-99-7227-2_3

as the number of features in a dataset, then there are 2^d distinct subsets of features. Given that checking all possible feature subsets in a dataset is a time-consuming and impractical process in cases where we have a large feature space, researchers have sought to provide methods that estimate the best subset of features in a reasonable time during recent decades. Therefore, providing effective methods to find the best subset of features [6, 7] is a great challenge. A lot of feature selection algorithms have been proposed based on different strategies so far [6–9].

Swarm intelligence methods are a class of population-based metaheuristic algorithms used to estimate near-optimal solutions to complex optimization problems. These methods are based on the collective behavior of living organisms in nature that find their food or path. To solve optimization problems using these methods, the search space is simulated and the best solution is estimated based on the swarm intelligence of the organisms. Among the best known of these methods are the artificial bee colony optimization (ABC) algorithm based on the behavior of bees, the ant colony optimization (ACO) algorithm based on the behavior of ants, and the particle swarm optimization (PSO) algorithm based on the behavior of birds and fish. Swarm intelligence methods have been used in many feature selection applications. One of the approaches that has proven to be effective in this area is the ACO [10].

ACO simulates the behavior of real ants in nature to optimize NP-hard problems. A chemical called pheromone is the main component of the ACO algorithm, which is used for possible scanning of the search space and is based on random search. An ant colony tries to find the best food sources by exploring the possible paths in the search space. The ants communicate with each other using pheromone values and find the best routes through cooperation [11, 12]. To solve various optimization problems, the ACO algorithm has been used so far in machine learning and has performed very well in solving these tasks. One of these tasks is feature selection, and so far, several ACO-based feature selection methods have been proposed.

So far, ACO-based feature selection methods have been reviewed in several articles. However, some issues have not yet been addressed in these articles. One of the most important issues in presenting a feature selection method is its compatibility with a particular type of dataset. Some datasets have one or more class labels for each training sample (supervised). In some datasets, only some instances have labels (semisupervised); in others, there are no labels (unsupervised). Therefore, separate methods are provided for each of these data types, since the nature of the data is different. In reviewing several articles that have studied ACO-based feature selection methods, we found that none of these methods have been studied in terms of their application to different data.

The second category of feature selection methods is based on their interaction with the classifier (filter, wrapper, embedded, hybrid). This type of classification has been studied in previous review articles. However, many articles were omitted from these articles. Therefore, in this chapter, we will provide a comprehensive overview of ACO-based feature selection methods, with our main focus on reviewing the idea and nature of each method in two general categories of feature selection. Methods not covered in previous review articles or published later are also included in this chapter. From this comprehensive review, we can see that the ACO algorithm has been used less or not at all in which feature selection tasks, and we will also present what can be done in the future.

This chapter includes seven sections: After the introduction, in Sect. 2, related works in this field are discussed. Section 3 explains the ACO algorithm, its formulation, and its advantages. Section 4 describes the feature selection task and its variants based on different perspectives. In Sect. 5, the ACO-based feature selection algorithms are reviewed. Sections 6 and 7 provide a discussion and the conclusion of the chapter.

2 Related Works

In this section, we discuss works that have investigated ACO-based feature selection methods. Nayar et al. [13] have presented a work that investigates ACO-based feature selection. First, this work explains the ACO algorithm and its various improved versions. Then, the applications of the ACO algorithm in various real-world applications are discussed. Several ACO-based feature selection algorithms are briefly described and the existing challenges are listed. However, no specific analysis of feature selection algorithms, for what kind of data they are suitable, and what strategy they use to explore the search space is done in this article. Moreover, only some of the available methods are presented in the article.

Rostami et al. [14] provide a survey on swarm intelligence-based feature selection methods. In this article, due to the fact that the ACO-based methods have not explicitly been covered, only a small number of existing methods have been reviewed. However, these methods have been classified in terms of interaction with the learning algorithm, and the number of their goals in the feature selection process has also been examined.

Nguyen et al. [10] present a comprehensive survey of feature selection methods based on swarm intelligence. It can be said that this article is the most comprehensive review that has been done so far on ACO-based feature selection methods. This article aims to examine the methods in terms of exploring the search space and how to interact with the learning algorithm. However, due to the publication of this article in 2020, the methods published after that are not included.

Khurma et al. [15] provide a review of nature-inspired methods for feature selection applications. The main focus of this paper is to combine nature-inspired algorithms and describe this combination. In addition, it focuses on examining the methods presented in various real-world applications such as medical, microarrays, image processing, handwriting recognition, name recognition, etc. Therefore, They did not focus specifically on reviewing ACO-based feature selection methods and only reviewed them from a practical point of view.

Based on the above, it can be said that ACO-based feature selection methods have not yet been fully and currently presented in a specialized format for feature selection, and our goal in this chapter is to do so.

3 Ant Colony Optimization (ACO) Algorithm

Dorigo et al. [16] proposed the ACO algorithm to find an estimation of the optimal solutions in complicated optimization problems. ACO is a metaheuristic method of random search and swarm intelligence that simulates the behavior of a colony of ants in nature. The pheromone is the main part of the ACO algorithm that determines the quality of the solutions in the search space. An ant deposits a sequence of pheromones when it moves from the nest to the meal. The ants will not be trapped in a local optimum by evaporating the pheromone's value over time. Therefore, the selection probability of an optimal solution is higher when more ants are traveling on a path and the pheromone value is more intense.

Artificial ants have more capabilities in ACO than real ants. Each ant has a memory that stores the previous activity's history and evaluates the quality of the solution generated by each ant. To obtain better solutions, they update the pheromone trail globally and locally. Additionally, some heuristic information can be applied by ACO to explore the search space more effectively [17].

After initializing the parameters, the main process of the ACO algorithm is as follows: The ants travel through the search space and explore various solutions by adding pheromone trails over the path. The pheromone content evaporates uniformly at all points as the ants complete their solution. Ants update the pheromone values on their path [13]. The basic flow of the ACO algorithm is given in Fig. 1.

The problem must be formulated in an appropriate way to represent the ACO-based feature problem. The features can be illustrated as nodes of a graph, and the edges represent the relationships between the features. The optimal solution can be achieved after completing all iterations [13].

Suppose a graph G with a set of features $F = \{f_1, f_2, \ldots, f_m\}$ as graph nodes. This graph is simulated in Fig. 2 with six features. In the beginning, m ants accidentally begin to move on to d features of the graph with some initial pheromone values. In other words, every ant moves on graph G using a probabilistic rule. Then, ants update the pheromone values once the trip is completed according to the global updating rule [13].

A rule selects the next node in the graph. Ant k uses the following relation to travel from feature i to feature j:

$$p_{ij}^{k}(t) = \begin{cases} [\tau_i(t)][\eta(F_i, F_j)]^{\beta} \Big/ \sum_{u \in N_i [\tau_u][\eta(F_i, F_u)]^{\beta}}, \forall j \in N_i^k, \text{ if } q > q_0 \\ 0 \qquad\qquad\qquad\qquad\qquad\qquad\qquad\qquad , \text{otherwise} \end{cases} \quad (1)$$

where $p_{ij}^{k}(t)$ is a transition possibility to travel from feature i to feature j at time t for the k-*th* ant, τ_i is the pheromone value assigned to feature i, and $\eta(F_i, F_j)$ is the heuristic information between features (F_i, F_j). N_i^k is the set of neighbor features for feature i, q is a random number in the range [0, 1], and q_0 is a fixed variable ($0 \le q_0 \le 1$). The value of the parameter β is between [0, 1], specifying the pheromone amount and the relative significance of heuristic information. If $\beta = 0$,

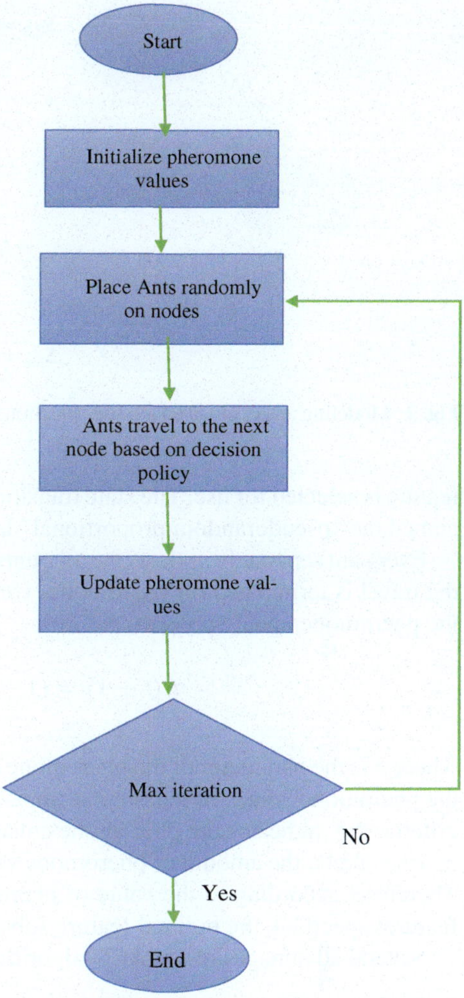

Fig. 1 Basic flow of the ACO algorithm

the decision is constructed according to the previous action history, and the heuristic information is neglected [13].

Each ant uses the following equation to travel from one feature to another:

$$i = \underset{u \in N^k}{\operatorname{argmax}} \left\{ [\tau_u][\eta(F_i, F_u)]^{\beta} \right\}, if q \leq q_0 \tag{2}$$

Parameters q and q_0 balance the exploration and exploitation abilities of the model. A random number q is produced every time an ant wants to move from feature i to the next feature. Each feature is chosen in proportion to its probability (Exploration) when $q > q_0$, utilizing Eq. 1. Otherwise, by Eq. 2 (Exploitation), the most beneficial

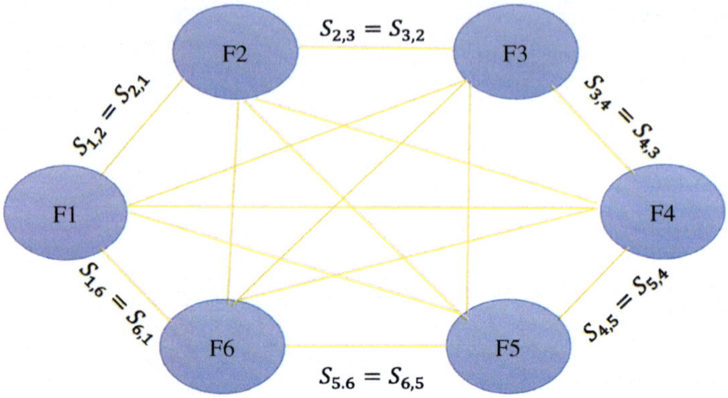

Fig. 2 Modeling an ACO-based feature selection problem using graph representation

feature is selected for use. The state transition rule is the composition of Eqs. 1 and 2, named the "pseudorandom proportional" law.

Every ant separately updates the amounts of the pheromone values in its route after the travel is terminated on graph nodes (features). The following formula conducts the pheromone update process:

$$\tau_i(t+1) = (1-\rho)\tau_i(t) + \Delta\tau_i \tag{3}$$

where ρ is the parameter of the pheromone decay rate and $\tau_i(t)$ and $\tau_i(t+1)$ indicate the pheromone values of feature i at times t and $t+1$, respectively. Based on some criteria, $\Delta\tau_i$ indicates the pheromone enhancement amount for feature i [12].

Eventually, the amount of pheromone value is the measure of fitness for features. Therefore, according to the value of pheromones, by selecting the m ($m > d$) top features specified, the optimal feature subset is determined [13].

Several advantages of the ACO algorithm are mentioned including:

- Performs calculations in a distributed manner.
- If a node fails, it has the ability to automatically change the shortest path.
- It is fast and converges quickly.
- It is simple and has high search power.
- It is robust.

4 Feature Selection

Preprocessing is an important step in extracting knowledge from raw data, which is done in two ways: feature extraction and feature selection. Feature selection is about finding optimal features, and feature extraction is about representing a combination of original features using transfer methods. In feature extraction, each new feature can be representative of one or more features. Feature selection, therefore, refers to searching for a subset of features to select relevant features and remove redundant or irrelevant features. Unlike feature extraction, feature selection leaves the data value unchanged despite dimensionality reduction. The primary objectives of feature selection are to improve the performance of a model through predictive accuracy, to improve the interpretability of the model, and to develop a robust model against overfitting to allow for better generalization [14, 18].

Feature selection can reduce model complexity and training time in machine learning. Therefore, the usage of feature selection has been extensively studied in machine learning tasks. These tasks include classification, regression, and clustering in various real-world applications. One of the goals of feature selection in these studies was to combat data volume due to the lack of scalability. Not all features in a dataset have the same discriminative power, and some contain redundant and irrelevant features. The increase in data volume caused the emergence of high-dimensional feature space. This phenomenon is known as the curse of dimensions, which is the exponential growth of data volume due to the increasing number of dimensions or attributes. Therefore, researchers have extensively studied feature selection methods to propose effective feature selection methods for high-dimensional data [9, 19].

Based on the interaction with the learning algorithms, embedded, wrapper, filter, and hybrid are four categories of feature selection methods. Filter methods evaluate features according to statistical criteria without using any classifier. Wrapper methods, on the other hand, use a learner to assess the predictive power of feature subsets. This means that different subsets of features are explored in a learning process to discover a subset with the highest predictive power. Embedded methods try to benefit from the advantages of two wrapper and filter techniques. In embedded methods, a learning algorithm, such as lasso, ridge regression, or autoencoder, initially runs over the original features to determine the predictive power of the features. The feature evaluation phase is performed based on these values. A hybrid method is a fusion of three previously mentioned categories, such as a combination filter and wrapper [20–22].

Unsupervised, supervised, and semisupervised are three groups of feature selection methods considering the existence of class labels. The degree of relevancy of each feature against the class label is the main measure of selecting features in supervised feature selection methods. Unsupervised feature selection methods handle datasets without any label information. The labeled samples are a small percentage of data for semisupervised data [23, 24].

For semisupervised and supervised data, there is not always one class label. Rather, in some datasets, multiple class labels exist. In a single-label dataset, all the samples have one class label; in a multi-label dataset, there are multiple labels. Single-label data are common in various applications today. For example, consider a news item. This news item can be categorized as political, economic, and social at the same time. Therefore, this message has three class labels. Therefore, there can be a different categorization of feature selection methods according to the number of labels: single-label and multi-label [25].

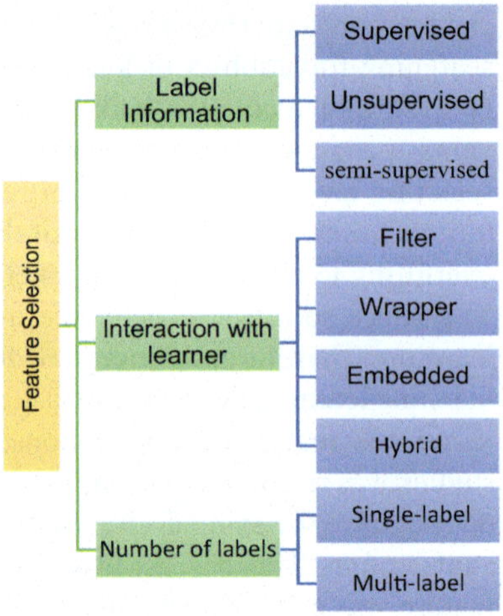

Fig. 3 General categorization of feature selection methods

In general, three categories of feature selection methods from different perspectives are presented in Fig. 3.

5 ACO-Based Feature Selection Algorithms

ACO-based feature selection methods are reviewed in this section from the supervision perspective (supervised, unsupervised, semisupervised). Moreover, in a separate subsection, we classify these methods according to the usage of the learning algorithm (filter, wrapper, embedded, and hybrid).

5.1 Supervised ACO-Based Feature Selection Methods

This section discusses supervised ACO-based feature selection methods for single-label and multi-label datasets separately.

5.1.1 Supervised Single-Label Methods

Gu-MOACOFS [26] has used the multiobjective ACO algorithm using Gmean and F_1 as its objectives for feature selection in imbalanced datasets. In this article, first, the data are partitioned into several partitions of samples using the bootstrap method, and then the feature selection method is applied to each group for the classification process.

ACOFS [27] is a hybrid feature selection method that utilizes a neural network as the classification estimator in the wrapper phase. In this method, the information gain of all features is calculated in the filter phase. Then, k ants travel on the feature graph using the information gain of features to find k different subsets of features. A neural network evaluates these subsets to save the local and global bests in each iteration. Additionally, the heuristic measurement and pheromone update rules are altered in this version of ACO.

RSFSACO [28] used the ACO algorithm and the rough set approach for feature selection. The heuristic value is first determined using the mutual information between each feature and the class label. In this method, unlike other ACO-based algorithms, where the ants are placed randomly on the nodes initially, all the ants start from a core. The core is used because the algorithm's convergence is performed faster, and the features around the center are selected faster. In the end, based on the concept of rough sets, the dimension of the feature space is reduced.

TSHFS-ACO [29] is proposed for high-dimensional data as a hybrid feature selection method. The redundancy between features is captured using the correlation coefficient metric as the heuristic information, and the pheromone values are initialized by the symmetrical uncertainty between features and the class label. In the TSHFS-ACO method, the interval strategy is applied to determine the number of selected features, and the K-nearest neighbor is the evaluator of features.

WFACOFS [30] is an ACO-based feature selection method considering a multiobjective optimization to assess the features. The amount of redundancy among features is captured by the cosine similarity metric. The features are initially selected based on this similarity in the filter stage. Then, the selected features are assessed in the wrapper stage using a neural network and K-nearest neighbor algorithms.

ABACO [31] is a feature selection method that utilizes a binary version of the ACO algorithm. In this method, each ant must preview all the graph nodes (features) and decide whether to select that feature at each stage. This decision is made by placing the feature in the classification process and checking its performance. Additionally, in this method, several feature evaluation criteria, including the correlation coefficient and f score, are considered to provide heuristic information.

ACO-S [32] is proposed for gene (feature) selection in microarray datasets. The information gain and Fisher score metrics are utilized to maintain heuristic information. An improved version of the ACO algorithm is used to determine the best features by the fuzzy logic control theory to tune the parameter values of the ACO.

ACOAR [33] uses the rough set theory to evaluate features. In this approach, the pheromone updating procedure is conducted on the graph nodes belonging to the best solution thus far. Additionally, it used a limit for pheromone values. A rapid procedure performs the solution construction phase.

ACO-En-LS [34] is a feature selection method proposed for text classification. The heuristic function of this method is defined according to local search and statistical metrics. It selects a definite number of features for each text category. Additionally, the feature subsets are assessed using the naïve Bayes classifier.

ASRank [35] utilized the ACO algorithm in a wrapper feature selection model to enhance pattern recognition system performance. In this method, unlike typical

ACO-based feature selection methods that use prior feature information as heuristic functions, the classification accuracy of feature subsets and the length of subsets are considered heuristics. Each ant must preview all the graph nodes (features) in each iteration.

Ant-EHFS [36] uses the ensemble of feature evaluation metrics to obtain the heuristic information of the ACO algorithm. In this method, multiple feature selection algorithms are applied to the feature space to evaluate features. A Type-I fuzzy is used to perform the ensemble feature selection process. The Euclidean distance between the features is also the second heuristic information used by the ACO algorithm.

Ant-MCDM [37] models the combination of multiple heuristics in the ACO algorithm as a process of multicriteria decision-making (MCDM). Instead of using one heuristic in the ACO algorithm, this method uses a combination of feature evaluation methods based on the redundancy between features and the relevance of each feature to the class label. Each of these evaluation criteria is considered an expert in this method. Finally, MCDM summarizes these opinions and provides a heuristic for the ACO algorithm.

DEFRFS and ACFRFS [38] are two hybrid ACO-based feature selection methods that have been proposed for microarray data to predict cancer. DEFRFS uses differential evolution with a fuzzy rough set, and ACFRFS uses a fuzzy rough set for multiobjective feature selection. In both methods, the fuzzy rough set is used as the fitness function in ACO. In DEFRFS, the fuzzy rough set is hybridized with differential evolution in ACO for the first time.

ACOFS [39] is proposed to reduce the dimensionality of data for image classification tasks. In the basic ACO algorithm and other improvements, the graph of the search space is usually complete. However, this method defines this graph as a directed graph to search the feature space faster and reduce the memory requirement. This problem is optimized using two metrics: classification performance and feature set size.

UFSACO [40] uses a fusion of the ACO algorithm and graph fusion for feature selection. The edges of the graph are weighted according to the similarity of the features. The feature space is assumed to be a clustered graph by utilizing a community detection algorithm. A filter metric, the separability index, is utilized to evaluate the predictive power of the features to the class label. Selecting an optimal subset of unique and relevant features is the main objective of this method.

5.1.2 Supervised Multi-label Methods

MLACO [41] is the first ACO-based multi-label feature selection method. The initial pheromone values are determined by the maximum cosine similarity between the set of features and labels. Additionally, the redundancy between the features and their relevance to the class labels is achieved by the correlation coefficient metric as heuristic information.

SMFSACO [42] uses mutual information between features to construct a complete graph for the search space, where the mutual information between two features is the weight of the edge between them. A combination of redundancy between features and relevance of features with the class labels is used according to the mutual information for the transition rule in ACO.

MLDGC [43] places similar features in a group using the k-nearest neighbor algorithm utilizing a density-based clustering approach. Then, the ACO algorithm is applied to choose the unique feature with high predictive power. In this version of ACO, the ants start in a cluster, and every time a random value is less than a predefined threshold, the ant stays in that cluster and examines only the features of that cluster. Otherwise, it can explore other feature clusters as well.

Ant-TD [44] is the first version of the ACO algorithm that employs a heuristic learning procedure. Due to the importance of heuristics in searching the feature space in the ACO algorithm, the ACO problem is modeled as a reinforcement learning procedure. In the sense that each feature represents a state, the ants' travel between features represents the action. In reinforcement learning, each action has a reward or a penalty. Therefore, the heuristic values change based on these rewards and punishments at each stage. This process of updating heuristics is done by temporal difference.

5.2 *Unsupervised ACO-Based Feature Selection Methods*

FSvACO [45] is a hybrid method to solve unsupervised feature selection problems. The cosine similarity between features is considered as heuristic information. After applying ACO, the pheromone values are used to sort the features in descending order, and eighty percent of them are selected for the next stage. Then, the features are fed to a learning algorithm by a forward feature selection approach to select the best subset.

UFSACO [46] is proposed to deal with unsupervised datasets without using any learning algorithms. In this method, the features are considered the nodes of a complete graph. The weight of edges between graph nodes is determined by the similarity between features. The ACO algorithm has been applied to this graph, and the features are sorted according to the pheromone values at the end.

FSBACOM [47] has modeled the feature space as a circular graph with weighted edges using the Pearson correlation coefficient between feature pairs. This modeling has been done due to the reduction of computational complexity compared to the complete graph, which reduces the processing time more swiftly, especially for high-dimensional datasets. Additionally, a mutation strategy is applied to avoid trapping in the local optimum.

UPFS [48] mapped features in a complete graph and the Pearson correlation coefficient obtained the heuristic information. In each iteration, when an ant wants to travel from one node to another, the redundancy of that node is computed with all previously selected nodes. The node with the minimum redundancy is chosen. The features are ordered based on the pheromone values at the end.

Table 1 Classifying ACO-based feature selection methods according to the interaction with the learner

Category	ACO-based feature selection methods
Filter	SemiACO [13], UPFS [48], FSBACOM [47], UFSACO [46], FSvACO [45], MLACO [41], Ant-TD [44], MLDGC [43], SMFSACO [42], Ant-EHFS [36], Ant-MCDM [37], RSFSACO [28], UFSACO [40], ACOAR [33]
Wrapper	Gu-MOACOFS [26], ABACO [31], ACO-S [32], ACO-En-LS [34], ASRank [35], ACOFS [39]
Embedded	–
Hybrid	ACOFS [27], TSHFS-ACO [29], WFACOFS [30], DEFRFS [38], ACFRFS [38]

5.3 *Semisupervised ACO-Based Feature Selection Methods*

To date, only one ACO-based semisupervised feature selection algorithm has been presented. SemiACO [13] uses a nonlinear heuristic function, unlike most ACO-based feature selection methods. The maximum information coefficient (MIC) and the mutual information of the labeled samples are used to initialize the state vector and the pheromone values. The ACO algorithm is integrated with reinforcement learning to update the heuristic information in each iteration. Moreover, the initial heuristic value is a mixture of the relevance of the labeled samples and the redundancy of all samples.

5.4 *Classifying Based on the Interaction with the Learner*

In this section, we categorize the methods that were reviewed in the previous three sections based on how they interact with the learning algorithm into four categories: embedded, wrapper, filter, and hybrid. This classification is presented in Table 1.

6 Discussion

In the previous section, ACO-based feature selection methods were categorized from two viewpoints of feature selection. Based on this review, we found that most of the techniques presented thus far have been for supervised data, and other areas have received less attention. Only one method based on the ACO algorithm is presented for semisupervised feature selection, which is the filter type. Therefore, methods based on filter, wrapper, embedded, and hybrid approaches can still be considered in this field. Also, the ACO algorithm has not been used in semisupervised multi-label problem.

For unsupervised data, several papers are presented, which are all filters. Considering that there is no class label in this type of data, wrapper methods cannot be used. The noteworthy point is the lack of embedded approaches in all kinds of data, which can be an interesting challenge for researchers in this field. Perhaps using a learning algorithm such as lasso and its coefficients as heuristic values in the ACO algorithm is a simple solution to this problem, which seems not to have been considered so far.

In supervised multi-label data, all the presented methods are of the filter type, and wrapper and embedded methods have not been investigated so far, which can be an untouched space for research. The issues discussed in this section are analyzing feature selection methods for data gathered already and not being updated over time. However, in many new real-world applications, features and instances can be added to the data in a streaming manner. Feature selection for such data is also a necessary task known as online feature selection. It should be noted that no method based on the ACO algorithm has been used in any type of online data, such as supervised, semi-supervised, or unsupervised.

In addition, ACO-based methods have not been applied so far in cost-based feature selection, which is a feature selection task in which the cost of feature production is also involved in their evaluation. Moreover, the ACO-based methods are not applied in another task called Multiview feature selection, where the features are gathered from multiple sources.

7 Conclusion

This chapter presents an overview of ACO-based feature selection methods. In this study, the usage of ACO-based feature selection methods is discussed in different data types. The proposed methods are reviewed based on the type of data provided to reduce its dimension. We have realized that the ACO algorithm has shown considerable results in feature selection tasks. However, there are still tasks in which the performance of the ACO algorithm has not been investigated, and these tasks were identified in this study. Therefore, it can be said that the methods based on the ant algorithm can perform well in the field of feature selection. This action can be done by not choosing simple heuristics. Methods such as combining different heuristics or using embedded approaches can be a solution to this problem.

References

1. Dhal P, Azad C (2022) A comprehensive survey on feature selection in the various fields of machine learning. Appl Intell 52:4543–4581. https://doi.org/10.1007/s10489-021-02550-9
2. Jia W, Sun M, Lian J, Hou S (2022) Feature dimensionality reduction: a review. Complex Intell Syst 8:2663–2693. https://doi.org/10.1007/s40747-021-00637-x

3. Hashemi A, Dowlatshahi MB, Nezamabadi-pour H (2021) Minimum redundancy maximum relevance ensemble feature selection: a bi-objective Pareto-based approach. J Soft Comput Inf Technol
4. Bayati H, Dowlatshahi MB, Hashemi A (2022) MSSL: a memetic-based sparse subspace learning algorithm for multi-label classification. Int J Mach Learn Cyber 13:3607–3624. https://doi.org/10.1007/s13042-022-01616-5
5. Miri M, Dowlatshahi MB, Hashemi A et al (2022) Ensemble feature selection for multi-label text classification: an intelligent order statistics approach. Int J Intell Syst 37:11319–11341. https://doi.org/10.1002/int.23044
6. Dowlatshahi MB, Hashemi A (2023) Unsupervised feature selection: a fuzzy multi-criteria decision-making approach 20:55–70
7. Hashemi A, Dowlatshahi MB, Nezamabadi-pour H (2020) MFS-MCDM: multi-label feature selection using multi-criteria decision making. Knowl-Based Syst 206:106365. https://doi.org/10.1016/j.knosys.2020.106365
8. Hashemi A, Bagher Dowlatshahi M, Nezamabadi-pour H (2021) An efficient Pareto-based feature selection algorithm for multi-label classification. Inf Sci 581:428–447. https://doi.org/10.1016/j.ins.2021.09.052
9. Hashemi A, Dowlatshahi MB, Nezamabadi-pour H (2022) Ensemble of feature selection algorithms: a multi-criteria decision-making approach. Int J Mach Learn Cyber 13:49–69. https://doi.org/10.1007/s13042-021-01347-z
10. Nguyen BH, Xue B, Zhang M (2020) A survey on swarm intelligence approaches to feature selection in data mining. Swarm Evol Comput 54:100663. https://doi.org/10.1016/j.swevo.2020.100663
11. Dey N, Ashour AS, Bhattacharyya S (2020) Applied nature-inspired computing: algorithms and case studies. Springer, Singapore
12. Karimi F, Dowlatshahi MB, Hashemi A (2023) SemiACO: a semi-supervised feature selection based on ant colony optimization. Expert Syst Appl 214:119130. https://doi.org/10.1016/j.eswa.2022.119130
13. Nayar N, Gautam S, Singh P, Mehta G (2021) Ant colony optimization: a review of literature and application in feature selection. In: Smys S, Balas VE, Kamel KA, Lafata P (eds) Inventive computation and information technologies. Springer, Singapore, pp 285–297
14. Rostami M, Berahmand K, Nasiri E, Forouzandeh S (2021) Review of swarm intelligence-based feature selection methods. Eng Appl Artif Intell 100:104210. https://doi.org/10.1016/j.engappai.2021.104210
15. Abu Khurma R, Aljarah I, Sharieh A et al (2022) A review of the modification strategies of the nature inspired algorithms for feature selection problem. Mathematics 10:464
16. Dorigo M, Birattari M, Stutzle T (2006) Ant colony optimization. IEEE Comput Intell Mag 1:28–39. https://doi.org/10.1109/MCI.2006.329691
17. Nayar N, Ahuja S, Jain S (2019) Swarm intelligence for feature selection: a review of literature and reflection on future challenges. In: Kolhe ML, Trivedi MC, Tiwari S, Singh VK (eds) Advances in data and information sciences. Springer, Singapore, pp 211–221
18. Dokeroglu T, Deniz A, Kiziloz HE (2022) A comprehensive survey on recent metaheuristics for feature selection. Neurocomputing 494:269–296. https://doi.org/10.1016/j.neucom.2022.04.083
19. Hashemi A, Dowlatshahi MB, Nezamabadi-pour H (2020) MGFS: a multi-label graph-based feature selection algorithm via PageRank centrality. Expert Syst Appl 142:113024. https://doi.org/10.1016/j.eswa.2019.113024
20. Dalvand A, Dowlatshahi MB, Hashemi A (2022) SGFS: a semi-supervised graph-based feature selection algorithm based on the PageRank algorithm. In: 2022 27th international computer conference, computer society of Iran (CSICC), pp 1–6
21. Hashemi A, Dowlatshahi MB (2022) An ensemble of feature selection algorithms using OWA operator. In: 2022 9th Iranian joint congress on fuzzy and intelligent systems (CFIS), pp 1–6
22. Hancer E, Xue B, Zhang M (2020) A survey on feature selection approaches for clustering. Artif Intell Rev 53:4519–4545. https://doi.org/10.1007/s10462-019-09800-w

23. Hashemi A, Dowlatshahi MB (2023) A fuzzy integral approach for ensembling unsupervised feature selection algorithms. In: 2023 28th international computer conference, computer society of Iran (CSICC), pp 1–6
24. Solorio-Fernández S, Carrasco-Ochoa JA, Martínez-Trinidad JF (2020) A review of unsupervised feature selection methods. Artif Intell Rev 53:907–948.https://doi.org/10.1007/s10462-019-09682-y
25. Kashef S, Nezamabadi-pour H, Nikpour B (2018) Multilabel feature selection: a comprehensive review and guiding experiments. Wiley Interdiscip Rev: Data Min Knowl Discov 8:e1240
26. Liu Y, Wang Y, Ren X et al (2019) A classification method based on feature selection for imbalanced data. IEEE Access 7:81794–81807. https://doi.org/10.1109/ACCESS.2019.2923846
27. Kabir MM, Shahjahan M, Murase K (2012) A new hybrid ant colony optimization algorithm for feature selection. Expert Syst Appl 39:3747–3763. https://doi.org/10.1016/j.eswa.2011.09.073
28. Chen Y, Miao D, Wang R (2010) A rough set approach to feature selection based on ant colony optimization. Pattern Recogn Lett 31:226–233. https://doi.org/10.1016/j.patrec.2009.10.013
29. Ma W, Zhou X, Zhu H et al (2021) A two-stage hybrid ant colony optimization for high-dimensional feature selection. Pattern Recogn 116:107933. https://doi.org/10.1016/j.patcog.2021.107933
30. Ghosh M, Guha R, Sarkar R, Abraham A (2020) A wrapper-filter feature selection technique based on ant colony optimization. Neural Comput Appl 32:7839–7857
31. Kashef S, Nezamabadi-pour H (2015) An advanced ACO algorithm for feature subset selection. Neurocomputing 147:271–279. https://doi.org/10.1016/j.neucom.2014.06.067
32. Li Y, Wang G, Chen H et al (2013) An ant colony optimization based dimension reduction method for high-dimensional datasets. J Bionic Eng 10:231–241. https://doi.org/10.1016/S1672-6529(13)60219-X
33. Ke L, Feng Z, Ren Z (2008) An efficient ant colony optimization approach to attribute reduction in rough set theory. Pattern Recogn Lett 29:1351–1357. https://doi.org/10.1016/j.patrec.2008.02.006
34. Janaki Meena M, Chandran KR, Karthik A, Vijay Samuel A (2012) An enhanced ACO algorithm to select features for text categorization and its parallelization. Expert Syst Appl 39:5861–5871. https://doi.org/10.1016/j.eswa.2011.11.081
35. Kanan HR, Faez K (2008) An improved feature selection method based on ant colony optimization (ACO) evaluated on face recognition system. Appl Math Comput 205:716–725. https://doi.org/10.1016/j.amc.2008.05.115
36. Joodaki N, Dowlatshahi M, Joodaki M (2023) ANT-EHFS: ANT colony optimization equipped with an ensemble of heuristics through fuzzy logic for feature selection. J Mahani Math Res Cent 12
37. Hashemi A, Joodaki M, Joodaki NZ, Dowlatshahi MB (2022) Ant colony optimization equipped with an ensemble of heuristics through multi-criteria decision making: a case study in ensemble feature selection. Appl Soft Comput 124:109046. https://doi.org/10.1016/j.asoc.2022.109046
38. Meenachi L, Ramakrishnan S (2020) Differential evolution and ACO based global optimal feature selection with fuzzy rough set for cancer data classification. Soft Comput 24:18463–18475. https://doi.org/10.1007/s00500-020-05070-9
39. Chen B, Chen L, Chen Y (2013) Efficient ant colony optimization for image feature selection. Signal Process 93:1566–1576. https://doi.org/10.1016/j.sigpro.2012.10.022
40. Moradi P, Rostami M (2015) Integration of graph clustering with ant colony optimization for feature selection. Knowl-Based Syst 84:144–161. https://doi.org/10.1016/j.knosys.2015.04.007
41. Paniri M, Dowlatshahi MB, Nezamabadi-pour H (2020) MLACO: a multi-label feature selection algorithm based on ant colony optimization. Knowl-Based Syst 192:105285. https://doi.org/10.1016/j.knosys.2019.105285
42. Hatami M, Mehrmohammadi P, Moradi P (2020) A multi-label feature selection based on mutual information and ant colony optimization. IEEE, pp 1–6

43. Kakarash ZA, Mardukhia F, Moradi P (2023) Multi-label feature selection using density-based graph clustering and ant colony optimization. J Comput Des Eng 10:122–138
44. Paniri M, Dowlatshahi MB, Nezamabadi-pour H (2021) Ant-TD: ant colony optimization plus temporal difference reinforcement learning for multi-label feature selection. Swarm Evol Comput 64:100892. https://doi.org/10.1016/j.swevo.2021.100892
45. Eroglu DY, Akcan U (2023) Feature Selection via ACO
46. Tabakhi S, Moradi P, Akhlaghian F (2014) An unsupervised feature selection algorithm based on ant colony optimization. Eng Appl Artif Intell 32:112–123. https://doi.org/10.1016/j.engappai.2014.03.007
47. Manbari Z, Akhlaghian Tab F, Salavati C (2019) Fast unsupervised feature selection based on the improved binary ant system and mutation strategy. Neural Comput Applic 31:4963–4982. https://doi.org/10.1007/s00521-018-03991-z
48. Dadaneh BZ, Markid HY, Zakerolhosseini A (2016) Unsupervised probabilistic feature selection using ant colony optimization. Expert Syst Appl 53:27–42. https://doi.org/10.1016/j.eswa.2016.01.021

Chapter 4
Diagnosis Support for Diabetes with Ant Colony Optimization

Maheswata Moharana, Fahmida Khan, and Subrat Kumar Pattanayak

1 Introduction

Diabetes is an assortment of metabolic disorders. It can be defined by hyperglycemia caused by deficiencies in insulin secretion, insulin action, or both. Numerous organs, especially the kidneys, heart, blood arteries and nerves, can suffer long-term damage, and failure as a result of diabetes-related chronic hyperglycemia. [1]. Almost every population in the world is affected by diabetes mellitus and other milder forms of glucose intolerance, including impaired glucose tolerance, and epidemiological data suggest that without strong prevention and control initiatives, the global prevalence of diabetes will continue to rise [2, 3]. In 2010, it was estimated that 285 million individuals globally, had suffered from diabetes [4]. Similarly, according to the International Diabetes Federation's (IDF) 9th edition report, there are approximately 351.7 million people aged between 20 and 64 years worldwide who have diabetes, diagnosed or undiagnosed, and approximately 90% of them suffer from type 2 diabetes. By 2030, there will be 417.3 million individuals who suffer from diabetes, and by 2045, there will be 486.1 million diabetes cases [5]. Most cases are going in developing and middle-income nations. To better prevent and limit its incidence, diabetes mellitus must now be identified early and treated as a serious worldwide health concern. There are numerous pathogenic pathways that lead to the development of diabetes. These range from abnormalities that cause insulin to be resistant to it to auto-immune damage to the pancreatic beta-cells that result in insulin deficiency. Diabetes impairments in protein, lipid, and carbohydrate metabolism are caused by inadequate insulin action on target tissues. Insufficient insulin secretion and/or reduced insulin responses at one or more points in the intricate hormonal action pathways cause insufficient

M. Moharana (✉) · F. Khan · S. K. Pattanayak (✉)
Department of Chemistry, National Institute of Technology Raipur, Raipur 492010, India
e-mail: moharana.sweta@gmail.com

S. K. Pattanayak
e-mail: skpiitbbs@gmail.com

N. Dey (ed.), *Applications of Ant Colony Optimization and its Variants*, Springer Tracts in Nature-Inspired Computing, https://doi.org/10.1007/978-981-99-7227-2_4

insulin action. Insulin secretion and action deficiencies sometimes occur in the same patient, and it is frequently difficult to determine which aberration, if either alone, is the main contributor to hyperglycemia [6]. A severe case of hyperglycemia is characterized by polyuria, polydipsia, and weight loss, occasionally accompanied by polyphagia and impaired eyesight. Chronic hyperglycemia can also be accompanied by growth impairment and an increased risk of contracting certain diseases. Hyperglycemia with ketoacidosis or nonketotic hyperosmolar syndrome is two immediate, potentially fatal effects of uncontrolled diabetes [7].

Diabetes can be categorized into four categories based on its path physiology: type 1 and 2 diabetes and genetic diabetes [8, 9]. Approximately 5–10% of every case of diabetes is type 1 diabetes. However, type 2 diabetes, which affects more than 90% of individuals, causes microvascular and macrovascular consequences that are extremely distressing for careers and patients and place an enormous burden on healthcare systems [10]. Type 2 diabetes is a multifactorial, complex, and chronic metabolic condition that is influenced by a person's age lifestyle, diet, genetics, and environment. The stage of type 2 diabetes may be asymptomatic and remain unnoticed for several years [11]. The current rise in diabetes cases can be attributed to people's lack of awareness of their health status, which raises the probability that they have type 2 diabetes as well as their ignorance leads to various complications such as renal diseases, heart disease, brain damage, and abdominal injury. To ensure that patients are made aware of the issue and to minimize the possibility of serious problems, it is crucial to undertake an early diagnosis of type 2 diabetes.

Data miming can help the medical field in making treatment predictions and diagnoses. Data mining was used due to the fact that significant volumes of data were involved. Employing data mining techniques, patterns are extracted can reveal the pertinent information. The information obtained from decision-making can be aided better by data mining, and improve working efficacy [12]. There have been numerous attempts to diagnose diabetes based on their symptoms; however, this strategy has produced false assumptions and hasn't produced good criteria for creating an efficient associative classifier. Hence it was reported the use of optimization and classification techniques along with association rule mining to increase the accuracy of diabetes diagnosis. Ant colony-based categorization method has been proposed for the purpose of capturing a set of fuzzy guidelines for diabetes diagnosis [13]. The hard-combinatorial optimization problem is solved by using this metaheuristic technique called the ACO algorithm. An example of utilizing ant colony optimization is to anticipate disease in the area of medical diagnosis. For example, this technique was utilized to predict the tumor at the molecular level [14]. In addition, it can also be used to diagnose heart disease [15].

The present chapter discusses about various traditional methods along with current artificial intelligent-based approaches for diabetes diagnosis with the special focus on classification models using the implementation of ant colony optimization algorithms.

2 Diagnostic Approaches for Diabetes

The diagnosis of diabetes in traditional Chinese medicine is based on data gathered through four types of diagnostic procedures: looking, hearing, touching, and smelling. The two most frequent procedures are checking the tongue and taking the pulse. Studies show a moderately significant link between tongue images and diabetes in traditional Chinese medicine [16–18]. In earlier times, doctors used testing, description, and personal experience to diagnose diseases and detect changes. Therefore, the conditions and doctor's own professional abilities might readily influence the outcomes. The accuracy of pattern recognition and the repeatability of conformation are uncertain in the absence of rigorous evaluation criteria [19]. The diagnosis of diabetes requires the confirmation of chronic hyperglycemia. Generally, three types of glycemia exist: the diabetic type, the normal type, and the borderline type. A fasting plasma glucose level of 7.0 mmol/l (126 mg/dl) or higher and/or a plasma glucose level of 11.1 mmol (200 mg/dl) or higher are both considered to be diabetic. Fasting plasma glucose must be below 6–1 mmol (110 mg/dl) and 2 h after plasma glucose must be below 7.8 mmol (140 mg/dl) to be considered normal type. Those who are neither normal types nor diabetic are considered borderline types [20, 21]. The following are different methods for clinical diagnosis [22]:

- Hyperglycemia meets the criteria for the "diabetic type" and is observed on two different days and by reconfirmation, diabetes mellitus is diagnosed.
- If one of the following three circumstances coexists, it can be diagnosed by a single plasma glucose test of the "diabetic type": (a) typical symptoms of diabetes mellitus; (b) HbA1c $\geq$ 6.5% using a standardized method; and (c) unmistakable diabetic retinopathy.
- It is advised to retest for PG after an adequate period if the diagnosis of diabetes cannot be confirmed by these approaches.
- In addition to determining whether diabetes is present or absent, the doctor also determines its etiology, glycemia stage, and whether any associated diseases or complications are present.

Like different sectors, the healthcare sector has benefited greatly from the development of numerous cutting-edge data mining algorithms and data analysis methodologies in the last few decades. In the field of medicine, data mining technology has become a vital tool for applications including disease prediction, diagnosis support, brain tumors, and breast cancer detection and treatment [23–27]. Data mining technology analyzes a significant volume of diagnostic medical data to find the information patterns hidden by diseases [28]. As a result, it allows for precise executive in the diagnosis and treatment of diseases. The identification of diabetes via data mining technology is an essential and intriguing research challenge as the volume and complexity of medical data rises over time. There were approximately 318,000 mobile health applications available to users worldwide in 2017 for managing diabetes on mobile devices with smartphones, smart watches, and tablets [29]. Different diabetes-related applications track different metrics, such as blood glucose readings, dietary intake and sugar, physical activity and weight tracking, information sharing with health and social professionals, and patient data.

Different diabetic applications track different metrics, like dietary intake, blood glucose readings, physical activity, weight tracking, information sharing with health. However, the majority of these already available mobile health applications for diabetes are created for users who have already received a positive diagnosis of the disease and its associated factors. This work is focused on the use of machine learning algorithms for the early identification of diabetes-related diseases [30]. Current medical research and doctors believe that early diagnosis of diseases will improve recovery prospects. With technological advancements, machine learning (ML) can be used to study diseases and diagnose illness at the very early stage [31]. Machine learning is a data-driven tool described as the essential component of artificial intelligence (AI) used for analyzing data [32, 33]. The intended approximate outcome is usually achieved when ML uses the dataset to determine the underlying pattern. There are a number of machine-learning-based approaches for assessing diabetes, including diagnosis, prediction, and complication analysis. As input to the machine learning algorithms for diabetes diagnosis, researchers used a patient's diabetes history and physical examination results, including diastolic blood pressure, body mass index, plasma glucose concentration, diet, weight, insulin, water consumption, blood pressure [34–36]. Support vector machine (SVM), decision trees, naïve Bayes (NB), k-nearest neighbor (kNN), and tree boosting techniques such as AdaBoost, XGBoost, and random forest (RF) are the most widely used machine learning algorithms [37–39]. Compared to ensemble algorithms such as random forest, AdaBoost, and XGBoost conventional algorithms such as SVM, kNN, NB, etc., perform poorly and have low accuracy levels. Because these ensemble learners are defined by a set of hyperparameters, a reliable classification model is developed for them in solving a global optimization problem [40]. Predictive biomarkers associated with diabetes-related diseases have been extensively studied in recent years. Triglycerides [41], glycated hemoglobin A1c (HbA1c) [42], fasting plasma glucose [43], adiponectin, and inflammatory markers [44] are among the recognized biomarkers for the detection of type II diabetes. The majority of these, however, do not have an effective potential to predict insulin resistance and prediabetes earlier. As a result, significant progress has been made in identifying abnormalities in the serum metabolome of obese, prediabetic, and type II diabetes patients. This has been done as an effort to find specific metabolite biomarkers that can reflect physiological abnormalities in the early stage of the disease more quickly. It is obvious that changes in a variety of blood metabolites accompany the several stages of the development of type II diabetes and may have some predictive value for this process, even though there is no single blood metabolite that truly fulfills the role of a biomarker [45]. Phospholipids and branched chain amino acids, including valine, leucine, and isoleucine, are some of the most widely used biomarkers for the progression of type II diabetes [46, 47]. Furthermore, plasma phospholipids such as phosphatidylinositol and sphingomyelin can distinguish between healthy people and people with type II diabetes [48]. Small, noncoding RNAs called microRNAs, which are important in posttranscriptional gene expression, have also been studied in prediabetes. MiR-192 and miR-193b were the significant microRNAs, among others. MiR-193b is crucial for the differentiation of brown adipocytes and the control of inflammation, while miR-192 regulates the tumor protein p53. Other significantly increased miRNAs in type II diabetes are miR-29a, miR-30d, miR-9, miR-123a2, miR-34a, and miR375 are assumed to be important for β-cell function. However, it was discovered that certain miRNAs inhibited the expression, generation, or release of insulin [49].

The existing medical tests for disease diagnosis are generally stressful and expensive for people. Hence, automation of medical tests is highly recommended nowadays. To make the diagnosis process faster and easier, many algorithm-based techniques were designed to make the automation process smoother. Recent decades have seen a rise in the use of decision support systems in the field of medical diagnosis. Researchers are more interested in designing medical expert systems all over the world. Computer-aided diagnosis (CAD) approaches are used by automated systems to predict diseases based on specific parameters. A valuable assessment of such medical data is provided by pattern recognition and data mining. Categorization methods are the most commonly used data mining methodology for making decisions from real-world data [50]. The achievement of the system may be impacted by the use of data; therefore, choosing physiological characteristics or attributes may have a greater impact on performance and outcomes. The diagnosis system's ability to anticipate outcomes accurately will be more impacted by the optimal feature selection, which may also result in saving time. There are different computation-based approaches, such as support vector machines, artificial neural networks, fuzzy systems, rough sets for the diagnosis of diabetes. The neural network is one of the different data mining analytical technologies that can be used to predict medical data. Backpropagation in neural networks (BPNs) employs a gradient-based method that either trains slowly or develops structures with local minima [51]. To find the network weights, one can employ widely used optimization techniques such as genetic algorithm (GA) [52], particle swarm optimization (PSO) [53], and ant colony optimization [54] in place of gradient-based learning techniques. A backpropagation network (BPN) has been used with GA, a stochastic general search method, to determine the various parameters, including the number of hidden nodes and hidden layers, the selection of relevant feature subsets, the learning rate, the momentum, and the initialization and optimization of network connection weights [55]. Genetics and natural selection were the concepts behind the optimization process of the genetic algorithm. The three main operators of a genetic algorithm are reproduction, crossover, and mutation [56]. Genetic algorithms are parallel in nature. They can simultaneously explore different pathways in the solution space since they can produce many offspring. If one path has achieved its conclusion, then eliminate it and continue working on other paths, giving them a better chance to each run of discovering the ideal response [57]. A population of randomly chosen chromosomes serves as the foundation for the implementation of the genetic algorithm. A population expressing the search space for solution to the problem as individuals is used in the iterative genetic algorithm approach. Each individual is represented by a genome, which is a finite string of symbols. An initial population of individuals is generated at random or heuristically in the basic genetic algorithm, and at each evolutionary step (generation step), the individuals in the current population are decoded and assessed in accordance with a fitness function that characterizes the optimization problem in the search space. To create a new population, people are chosen based on their fitness [58]. This ensures that the anticipated frequency of selection of an individual approximately corresponds to its relative performance in the population. As a result, high-fitness people have a better probability of reproducing and adding to the population than low-fit individuals. In this procedure, optimization problems are solved using strategies derived from natural evolution, such as inheritance, mutation, selection, and crossover, in a manner that resembles natural selection [50].

To produce the next generation from the existing population, the genetic algorithm applies one of three basic types of rules at each stage [59]:

- The selection rules choose individuals, referred to as parents, who will contribute to the population in the next generation.
- Crossover rules combine two parents to produce the next generation's offspring.
- Mutation rules arbitrarily alter each parent to produce offspring.

A swarm-based intelligent stochastic search technique called particle swarm optimization (PSO) was inspired by the natural way that bees swarm when they are looking for food. As a result, due to the flexibility of numerical experimentation, PSO has mostly been used to tackle a variety of optimization challenges [60]. The technique was developed by Kennedy and Eberhart [61–63] as an evolutionary computing method. Social behaviors such as fish schooling and bird flocking promote PSO techniques. The PSO evolutionary technique is used to train a multilayer perception neural network (MLPNN) and determine the best weight values for the trained network in a hybrid model known as particle swarm optimization with multilayer perception neural networks. The dataset was run using SVM algorithm to determine the best features for predicting the different forms of diabetes. To appropriately assess the used models, the datasets were subjected to twofold and fourfold cross-validation in a hybrid model known as PSO-MLPNNs. The performance of the model was then assessed using a variety of criteria, including recall, specificity, and total accuracy [64]. Despite the fact that PSO approaches are often used for disease diagnosis, there is still a gap in comparative studies.

3 Ant Colony Optimization (ACO) Algorithms as Diagnostic Support for Diabetes

An early and accurate identification of a disease is essential for effective therapy. Additionally, social well-being and quality of life may be impacted by this. Several studies have been performed to develop an artificial intelligence-based and decision support system-based computer-aided diagnostic system [65]. Since some diseases have widespread symptoms, the diagnostic procedure can occasionally be difficult, and optimization approaches are needed to solve the issue. The use of metaheuristic and optimization algorithms is common in intelligent systems that aim to improve classification accuracy [66]. The majority of metaheuristic algorithms take their cues from nature to deduce complicated correlations from simple and fundamental conditions. The best illustration of optimization is seen in nature. Each metaheuristic algorithm produces a starting population of solutions that are applied before it continuously advances from generation to generation in the direction of the ideal solution [67, 68].

In both industrial and scientific fields, optimization problems are vital.

Nurse time distribution scheduling, train scheduling, traveling, capacity planning, vehicle routing challenges, salesman problems, etc., are some real-world instances of optimization problems. For this reason, many optimization algorithms have been developed in recent years [69–72]. Ant colony optimization is one such algorithm, which is a probabilistic method for locating the best routes. To resolve various computational problems, the ant colony optimization algorithm is utilized in computer science and research [73]. Marco Dorigo first discussed ant colony optimization in his doctoral thesis in the 1990s [74]. This algorithm is based on how ants search for a path between their colony and a food source while foraging. First, it was applied to resolve the famous traveling salesman problems. Later, it was employed to resolve many challenging optimization issues. This probabilistic approach to computing problems may be simplified to identify good paths via graphs. Artificial ants are multiagent systems that mimic the behavior of real ants. The most prevalent paradigm is frequently the pheromone-based communication of ants. Numerous optimization problems requiring graphs, such as vehicle routing and internet routing, have found success using a combination of artificial ants and local search algorithms [75].

Some species of ants in the actual world roam randomly in search of food before returning to their colony while departure pheromone trails. If other ants discover such a road, they are more likely to trail it than to continue moving randomly, returning and re-establishing it if they eventually find food [76]. An artificial ant is a simple computational agent in this technique that looks for good solutions to an optimization problem. The optimization problem must be transformed into the problem of locating the shortest path on a weighted graph to use this optimization approach. Each ant begins by stochastically constructing a solution or the correct sequence for traveling the graph's edges. Second, comparisons are made between the routes taken by the various ants. The final step is to adjust the pheromone levels on each edge.

The fitness function of ant i at iteration t is defined as follows.

$$f_{(i)}(t) = \frac{1}{|A|} \sum_{j \in A} \frac{1}{1 + d_{ij}} \tag{3.1}$$

where A represents i's neighborhood, $|A|$ is the number of ants in A and d_{ij} is the distance between i and j.

The computational agent "Ant" in the ACO algorithm iteratively creates a solution to the current problem [77]. Solution states are denoted to as intermediate solutions.

An ant goes from state si to state sj, which corresponds to a more through intermediate solution, at any iteration. As a result, each ant k calculates a set Ak of viable expansions to its current state in each iteration and probabilistically advances to one of them. The probability of ant k moving from state si to an adjacent state sj at iteration t depends on the interaction of two values, such as the attractiveness of the move, and the pheromone trail of the move, Typically, k goes from state si to an adjacent state sj, as shown by

$$p_{ij}(t) = \frac{[\tau_{ij}(t)]^{\alpha}[n_{ij}]^{\beta}}{\sum_{Sj \in Ak}[(t)]^{\alpha}[\tau_{ij}(t)]^{\alpha}[n_{ij}]^{\beta}} \, if \; sj \in Ak \tag{3.2}$$

where τ_{ij} denotes the pheromone concentration along the route from i to j. n_{ij} signifies the value of the heuristic function, which describes the importance of choosing j. α and β determine the position of the pheromone and heuristic functions, respectively.

It would be mentioned that artificial ants have several unique abilities, including (i) memory, (ii) partial vision, and (iii) living in a discrete-time environment. ACO algorithms are also used to produce fuzzy classification rules known as FCS-ANTMINER [78]. To create a set of classification rules, ANTMINER used artificial ants. The results of ANTMINER were encouraging because it used a simpler function as heuristic information rather than an entropy parameter to minimize the computing overhead [79]. Utilizing the publicly accessible Pima Indian Diabetes dataset from the University of California, Irvine website [80], the proposed technique has been tested. According to the results, this algorithm is capable of detecting the diabetic illness signature with comparable or even higher correctness than the outcomes obtained by earlier works. As this technique generates a small number of fuzzy-if-then rules with modest lengths, this approach is easily understandable.

The following procedures are used to determine each fuzzy if-then rule's certainty grade once ACO generates a collection of rules.

Step 1: Determine whether each training pattern $x_p = (x_{p1}, x_{p2}, \ldots\ldots x_{pn})$ is compatible with the fuzzy if–then rule Rj using the following product operation.

$$\mu_j(x_p) = \mu_{j1}(x_{p1}) \ldots\ldots \times \mu_{jn}(x_{pn}),\ p = 1, 2, 3 \ldots\ldots m, \tag{3.3}$$

where $\mu_{ji}(x_{pi})$ is the membership function of the ith attribute of the pth pattern.

Step 2: Determine the relative total of the training pattern's compatibility grades using the fuzzy if–then rule Rj for each class.

$$\beta_{Class\,h}(R_j) = \sum_{x_p \in Class\,h} \mu_j(x_p)/N_{Class\,h} \quad h = 1, 2, \ldots, c \tag{3.4}$$

where N Class h is the total number of training patterns whose corresponding class is Class h and $\beta_{Classh}(R_j)$ is the sum of the compatibility grades of the training patterns in Class h with the fuzzy if–then rule Rj.

Step 3: Determining the grade of certainty CF_j.

$$CF_j = (\beta_{Class_{hj}}(R_j) - \overline{\beta}) / \sum_{h=1}^{c} Class_h(R_j) \tag{3.5}$$

$$\overline{\beta} = \sum_{h \neq hj} \beta_{Class_h}(R_j)/(c-1) \tag{3.6}$$

The codes for each fuzzy if-then rule are strings. The five linguistic values are represented as follows. 0: don't care (DC), 1: small (S), 2: medium small (MS), 3: medium (M), 4: medium large (ML), 5: large (L) [81].

Iyer et al. developed [82] diagnostic models for diabetes diagnosis, such as J48 (decision tree algorithm) and naïve Bayes. By using the cross-validation approach and the percentage split technique, the data were split into a training set and test set. The data for the training and test sets are prepared using 10-fold cross-validation. Following data preprocessing, the dataset is run through the J48 algorithm using WEKA (Java Toolkit for different Data Mining Techniques), at which point the data are classified as "tested-positive" or "tested-negative" based on the outcome of the decision tree that has been built.

The following is the algorithm for carrying out the procedure:

INPUT: Preprocessed dataset in CSV format for the Pima Indian Diabetes Database from the National Institute of Diabetes and Digestive and Kidney Diseases.

OUTPUT: J48 Decision Tree Predictive Model with Nave Bayes Prediction Results and leaf nodes that are either tested-positive or tested-negative.

METHODOLOGY

1. WEKA tools are used to preprocess the dataset. On the dataset, the following operations are carried out:

   i. Replacement of Missing Values and
   ii. Values Normalization

2. The feature selection process deletes certain groups of attributes from the processed dataset.
3. WEKA is used to upload the final processed dataset.
4. The naïve Bayes algorithm and J48 decision tree are employed.
5. Cross-validation and percentage split approaches are used to create models for the algorithms. Based on occurrences that had been accurately identified, both models were evaluated.

4 Application of ACO Algorithm

There are numerous applications for ant colony optimization algorithm.

Routing and network optimization: In computer networks, transportation networks, and logistics networks, routing can be improved using ant colonies. It can assist in locating the quickest routes and enhance the effectiveness of data transfer or the transportation of goods [83].

Resource allocation and scheduling:

These approaches can be used to optimize resource scheduling and allocation in a variety of contexts, including project management, workforce optimization, and production planning. To cut expenses and increase efficiency, it might be helpful to schedule work and distribute resources in the best possible way [84].

Traveling salesman issues:

The traveling salesman problem, which determines the quickest path to take between a starting city and a specified collection of cities, is particularly well-suited for ant colony optimization. To determine the best answer, the ants imitate the process of exploration and pheromone trail reinforcement [85].

In health care sectors:

Ant colony optimization algorithms have been widely used in health care sectors to solve various optimization problems such as nurse scheduling, patient routing, in medical images like brain MRI, chest X-rays, and can assist radiologists and clinicians to diagnose diseases and abnormalities more accurately and efficiently [86].

In drug discovery:

The ACO algorithms have been used to optimize the molecular docking in drug design [87]. It can be used to identify the conformational space of ligands and identify the most favorable binding sites on target proteins. By simulating the pheromone trail deposition and evaporation of ants, the algorithm can effectively search for the best docking poses and optimize binding affinities. The algorithm has also been employed to predict the tertiary structure of proteins [88, 89]. The sequence allotment problem in bioinformatics also solved by the ACO algorithm [90].

5 Challenges and Limitations of ACO Algorithms

Though the ant colony optimization method has been demonstrated to be successful in tackling a number of optimization problems, there is still needs for improvements in terms of quality of its solutions and the rate at which it converges. Some of the challenges that must be improved in the ant colony optimization algorithm are discussed below.

Parameter tuning: For optimum performance, ant colony optimization techniques require fine-tuning of a number of factors. The quantity of ants, pheromone update rate, and pheromone evaporation rate are some of these characteristics. Since no one set of ideal parameter values exists, tuning these parameters can be time-consuming and difficult [91].

Complexity: Ant colony optimization techniques are capable of becoming extremely complicated, especially when dealing with large and challenging problem areas. Multiple ants traveling through the problem space, updating probabilities, and taking actions based on local information are issues that the algorithms must handle [92].

Convergence to local optima: Ant colony optimization algorithms have to potential to convergence local optima rather the global optima [73]. Accordingly, the algorithm might come up with a less than ideal solution to a problem, given the status of the pheromone trails at the time. Methods like adding randomization to the algorithm can assist to solve this problem.

Limited information sharing: The pheromone trails are updated by ant colony optimization algorithms using local information exchange amongst the ants. This signifies that ants do not have a comprehensive understanding of the problem space and merely take into account the information they gather from their nearby ants.

6 Future Perspectives

Machine learning (ML) is a new tool that can be used to study diseases and make early diagnoses of illness. The most important aspect of artificial intelligence (AI) utilized for data analysis is machine learning, a data-driven technology. For chronic renal illness, an intelligent prediction and classification system using the ant colony-based optimization method and density-based feature selection has been designed [93]. In chemical production processes, fault identification is challenging due to the amount of data that needs to be analyzed and the lack of fault data. The fault features are then chosen using a unique quantum ant colony optimization approach that combines support vector machines with support vector machines to remove system disturbances and noise from the high levels of data [94]. The detection of rolling element bearing faults used the Improved ACO-Support Vector Machine technique. A more accurate recognition rate can be achieved with an improved ant colony optimization-support vector machine technique [95]. Ant colony optimization, a bioinspired advancement optimization technique, and ant colony optimization optimizers are used to gradually improve the classifier's parameter estimations to increase classification accuracy [96].

References

1. American Diabetes Association (2010) Diagnosis and classification of diabetes mellitus. Diabetes Care 33(Supplement 1):S62–S69
2. Katsarou A, Gudbjörnsdottir S, Rawshani A, Dabelea D, Bonifacio E, Anderson BJ, Jacobsen LM, Lernmark Å (2017) Type 1 diabetes mellitus. Nat Rev Dis Primers 3(1):1–17
3. Adachi T, Ohta H, Yamada H, Futenma A, Kato K, Hirano K (1992) Quantitative analysis of extracellular-superoxide dismutase in serum and urine by ELISA with monoclonal antibody. Clinicachimica acta 212(3):89–102
4. Behera SS, Ray RC (2016) Konjac glucomannan, a promising polysaccharide of Amorphophallus konjac K. Koch in health care. Int J Biol Macromol 92:942–956
5. Rufo DD, Debelee TG, Ibenthal A, Negera WG (2021) Diagnosis of diabetes mellitus using gradient boosting machine (LightGBM). Diagnostics 11(9):1714
6. James DE, Stöckli J, Birnbaum MJ (2021) The etiology and molecular landscape of insulin resistance. Nat Rev Mol Cell Biol 22(11):751–771
7. Flood T (2010) The many faces of type 2 diabetes. Postgrad Med 122(4):218–223
8. Canivell S, Gomis R (2014) Diagnosis and classification of autoimmune diabetes mellitus. Autoimmun Rev 13(4–5):403–407
9. Ortiz-Martínez M, González-González M, Martagón AJ, Hlavinka V, Willson RC, Rito-Palomares M (2022) Recent developments in biomarkers for diagnosis and screening of type 2 diabetes mellitus. Curr DiabRep 22(3):95–115
10. Chatterjee S, Khunti K, Davies MJ (2017) Type 2 diabetes. The Lancet 389(10085):2239–2251
11. Pillon NJ, Loos RJ, Marshall SM, Zierath JR (2021) Metabolic consequences of obesity and type 2diabetes: balancing genes and environment for personalized care. Cell 184(6):1530–1544
12. Salama KM, Abdelbar AM, Otero FE, Freitas AA (2013) Utilizing multiple pheromones in an ant-based algorithm for continuous-attribute classification rule discovery. Appl Soft Comput 13(1):667–675
13. Thippa Reddy G, Khare N (2016) FFBAT-optimized rule based fuzzy logic classifier for diabetes. Int J Eng Res Afr 24:137–152
14. Yu H, Gu G, Liu H, Shen J, Zhao J (2009) A modified ant colony optimization algorithm for tumor marker gene selection. Genomics Proteomics Bioinform 7(4):200–208
15. Ramo FM (2013) Diagnosis of heart disease based on ant colony algorithm. Int J Comput Sci Inf Secur 11(5):77. (and pap smears (Marinakis Y, Dounias G, 2008)
16. Xu J, Xu JT, Zhu YH (2014) Exploration on correlation between glycometabolism and digital tongue picture in diabetic patients. Shanghai J Tradit Chin Med 48(11):11–17
17. He H, Yan S, Yang L (2013) Correlation of tongue and laboratory parameters in 5930 cases of type 2 diabetes. J Tradit Chin Med 54(23):2031–2034
18. Hsu PC, Huang YC, Chiang JY, Chang HH, Liao PY, Lo LC (2016) The association between arterial stiffness and tongue manifestations of blood stasis in patients with type 2 diabetes. BMC Complement Altern Med 16:1–7
19. Zhang J, Xu J, Hu X, Chen Q, Tu L, Huang J, Cui J (2017) Diagnostic method of diabetes based on support vector machine and tongue images. BioMed Res Int
20. Kuzuya T, Nakagawa S, Satoh J, Kanazawa Y, Iwamoto Y, Kobayashi M, Nanjo K, Sasaki A, Seino Y, Ito C, Shima K, & Committee of the Japan Diabetes Society on the Diagnostic Criteria of Diabetes Mellitus (2002) Report of the Committee on the classification and diagnostic criteria of diabetes mellitus. Diabetes Res Clin Pract 55(1):65–85
21. Kosaka K, Kuzuya T, Hagura R, Yoshinaga H (1996) Insulin response to oral glucose load is consistently decreased in established non-insulin-dependent diabetes mellitus: the usefulness of decreased early insulin response as a predictor of non-insulin-dependent diabetes mellitus. Diabet Med 13:109–119
22. Seino Y, Nanjo K, Tajima N, Kadowaki T, Kashiwagi A, Araki E, Ito C, Inagaki N, Iwamoto Y, Kasuga M, Hanafusa T, Ueki K (2010) Report of the Committee on the classification and diagnostic criteria of diabetes mellitus: The Committee of the Japan Diabetes Society on the diagnostic criteria of diabetes mellitus

23. Bi WL, Hosny A, Schabath MB, Giger ML, Birkbak NJ, Mehrtash A, Allison T, Arnaout O, Abbosh C, Dunn IF, Mak RH, Aerts HJ (2019). Artificial intelligence in cancer imaging: clinical challenges and applications. CA: A Cancer J Clin 69(2):127–157
24. Noreen N, Palaniappan S, Qayyum A, Ahmad I, Imran M, Shoaib M (2020) A deep learning model based on concatenation approach for the diagnosis of brain tumor. IEEE Access 8:55135–55144
25. Nayak J, Favorskaya MN, Jain S, Naik B, Mishra M (2021) Advanced machine learning approaches in cancer prognosis. Springer International Publishing
26. Cuocolo R, Caruso M, Perillo T, Ugga L, Petretta M (2020) Machine learning in oncology: a clinical appraisal. Cancer Lett 481:55–62
27. Dlamini Z, Francies FZ, Hull R, Marima R (2020) Artificial intelligence (AI) and big data in cancer and precision oncology. Comput Struct Biotechnol J 18:2300–2311
28. Babu S, Vivek EM, Famina KP, Fida K, Aswathi P, Shanid M, Hena M (2017) Heart disease diagnosis using data mining technique. In: 2017 international conference of electronics, communication and aerospace technology (ICECA), vol 1. IEEE, pp 750–753. (2017, April)
29. Grewal D, Hulland J, Kopalle PK, Karahanna E (2020) The future of technology and marketing: a multidisciplinary perspective. J Acad Mark Sci 48:1–8
30. Zou Q, Qu K, Luo Y, Yin D, Ju Y, Tang H (2018) Predicting diabetes mellitus with machine learning techniques. Front Genet 9:515
31. Phasinam K, Mondal T, Novaliendry D, Yang CH, Dutta C, Shabaz M (2022) Analyzing the performance of machine learning techniques in disease prediction. J Food Qual
32. Santosh KC (2020) AI-driven tools for coronavirus outbreak: need of active learning and cross-population train/test models on multitudinal/multimodal data. J Med Syst 44:1–5
33. Dey N, Ashour AS, Bhattacharyya S (eds) (2020) Applied nature-inspired computing: algorithms and case studies. Springer Singapore
34. Karthik S, Sudha M (2018) A survey on machine learning approaches in gene expression classification in modeling computational diagnostic system for complex diseases. Int J Eng Adv Technol 8(2):182–191
35. Sarwar A, Ali M, Manhas J, Sharma V (2020) Diagnosis of diabetes type-II using hybrid machine learning based ensemble model. Int J Inf Technol 12:419–428
36. Ibrahim I, Abdulazeez A (2021) The role of machine learning algorithms for diagnosing diseases. J Appl Sci Technol Trends 2(01):10–19
37. Ahmed N, Ahammed R, Islam MM, Uddin MA, Akhter A, Talukder MA, Paul BK (2021) Machine learning based diabetes prediction and development of smart web application. Int J Cogn Comput Eng 2:229–241
38. Shahi TB, Sitaula C, Paudel N (2022) A hybrid feature extraction method for Nepali COVID-19-related tweets classification. Comput Intell Neurosci
39. Mithra Raj M, Arul Jothi JA (2022) Website phishing detection using machine learning classification algorithms. In: International conference on applied informatics.Springer International Publishing, Cham, pp 219–233. (2022, October)
40. Koc K, Ekmekcioğlu Ö, Gurgun AP (2021) Integrating feature engineering, genetic algorithm and tree-based machine learning methods to predict the postaccident disability status of construction workers. Autom Constr 131:103896
41. Olivarius N, Nielsen N, Andreasen A (2001) Diabetic retinopathy in newly diagnosed middle-aged and elderly diabetic patients. Prevalence and interrelationship with microalbuminuria and triglycerides. Graefes Arch Clin Exp Ophthalmol 239:664–672
42. Hu Y, Liu W, Chen Y, Zhang M, Wang L, Zhou H, Wu P, Teng X, Dong Y, Zhou JW, Jia Y (2010) Combined use of fasting plasma glucose and glycated hemoglobin A1c in the screening of diabetes and impaired glucose tolerance. Acta Diabetol 47:231–236
43. Trujillo J, Vigo A, Reichelt A, Duncan BB, Schmidt MI (2014) Fasting plasma glucose to avoid a full OGTT in the diagnosis of gestational diabetes. Diabetes Res Clin Pract 105(3):322–326
44. Schulze MB, Rimm EB, Shai I, Rifai N, Hu FB (2004) Relationship between adiponectin and glycemic control, blood lipids, and inflammatory markers in men with type 2 diabetes. Diabetes Care 27(7):1680–1687

45. Fikri AM, Smyth R, Kumar V, Al-Abadla Z, Abusnana S, Munday MR (2020) Prediagnostic biomarkers of type 2 diabetes identified in the UAE's obese national population using targeted metabolomics. Sci Rep 10(1):17616
46. Guasch-Ferré, M, Hruby A, Toledo E, Clish CB, Martínez-González MA, Salas-Salvadó J, Hu FB (2016) Metabolomics in prediabetes and diabetes: a systematic review and meta-analysis. Diabetes Care 39(5):833–846
47. Roberts LD, Koulman A, Griffin JL (2014) Toward metabolic biomarkers of insulin resistance and type 2 diabetes: progress from the metabolome. Lancet Diabetes Endocrinol 2(1):65–75
48. Long J, Yang Z, Wang L, Han Y, Peng C, Yan C, Yan D (2020) Metabolite biomarkers of type 2 diabetes mellitus and prediabetes: a systematic review and meta-analysis. BMC Endocr Disord 20(1):117
49. Dorcely B, Katz K, Jagannathan R, Chiang SS, Oluwadare B, Goldberg IJ, Bergman M (2017) Novel biomarkers for prediabetes, diabetes, and associated complications. Diabetes, metabolic syndrome and obesity: targets and therapy, pp 345–361
50. Alharbi A, Alghahtani M (2019) Using genetic algorithm and ELM neural networks for feature extraction and classification of type 2-diabetes mellitus. Appl Artif Intell 33(4):311–328
51. Nayak SC, Misra BB, Behera HS (2017) Artificial chemical reaction optimization of neural networks for efficient prediction of stock market indices. Ain Shams Eng J 8(3):371–390
52. Sapna S, Tamilarasi A, Pravin Kumar M (2012) Implementation of genetic algorithm in predicting diabetes. Int J Comput Sci 9:234–240
53. Wang D, Tan D, Liu L (2018) Particle swarm optimization algorithm: an overview. Soft Comput 22:387–408
54. Dorigo M, Birattari M, Stutzle T (2006) Ant colony optimization. IEEE Comput Intell Mag 1(4):28–39
55. Karegowda AG, Manjunath AS, Jayaram MA (2011) Application of genetic algorithm optimized neural network connection weights for medical diagnosis of pima Indians diabetes. Int J Soft Comput 2(2):15–23
56. Forrest S (1993) Genetic algorithms: principles of natural selection applied to computation. Science 261(5123):872–878
57. Verma G, Verma V (2012) Role and applications of genetic algorithm in data mining. Int J Comput Appl 48(17):5–8
58. Goldberg DE (1989) Genetic algorithms in search, optimization and machine learning. Addition-Westly, Reading MA
59. Ephzibah EP (2011) Cost effective approach on feature selection using genetic algorithms and fuzzy logic for diabetes diagnosis. arXiv:1103.0087
60. Nedjah N, de Oliveira PJA (2020) Simultaneous localization and mapping using swarm intelligence based methods. Expert Syst Appl 159:113547
61. Kennedy J, Eberhart R (1942) Particle swarm optimization In: Proceedings of IEEE international conference on neural networks. Perth, Australia, pp 1948, 1995
62. Kulkarni MNK, Patekar MS, Bhoskar MT, Kulkarni MO, Kakandikar GM, Nandedkar VM (2015) Particle swarm optimization applications to mechanical engineering-a review. Mater Today Proc 2(4–5):2631–2639
63. Valdez F, Vazquez JC, Melin P, Castillo O (2017) Comparative study of the use of fuzzy logic in improving particle swarm optimization variants for mathematical functions using coevolution. Appl Soft Comput 52:1070–1083
64. Qteat H, Awad M (2021) Using hybrid model of particle swarm optimization and multi-layer perceptron neural networks for classification of diabetes. Int J Intell Eng Syst 14(3)
65. Mirza S, Mittal S, Zaman M (2018) Decision support predictive model for prognosis of diabetes using SMOTE and decision tree. Int J Appl Eng Res 13(11):9277–9282
66. Ahmadi N (2020) Review of terrestrial and satellite networks based on machine learning techniques. J Soft Comput Decis Support Syst 7(3)
67. Boiroux D, Aradóttir TB, Nørgaard K, Poulsen NK, Madsen H, Jørgensen JB (2017) An adaptive nonlinear basal-bolus calculator for patients with type 1 diabetes. J Diabetes Sci Technol 11(1):29–36

68. Li X, Zhang J, Safara F (2021) Improving the accuracy of diabetes diagnosis applications through a hybrid feature selection algorithm. Neural Process Lett 1–17
69. Ernst AT, Jiang H, Krishnamoorthy M, Sier D (2004) Staff scheduling and rostering: a review of applications, methods and models. Eur J Oper Res 153(1):3–27
70. Qi J, Li S, Gao Y, Yang K, Liu P (2018) Joint optimization model for train scheduling and train stop planning with passengers distribution on railway corridors. J Oper Res Soc 69(4):556–570
71. Demirbilek M, Branke J, Strauss AK (2021) Home healthcare routing and scheduling of multiple nurses in a dynamic environment. Flex Serv Manuf J 33:253–280
72. Pratap S, Kumar MB, Saxena D, Tiwari MK (2016) Integrated scheduling of rake and stockyard management with ship berthing: a block based evolutionary algorithm. Int J Prod Res 54(14):4182–4204
73. Yang J, Zhuang Y (2010) An improved ant colony optimization algorithm for solving a complex combinatorial optimization problem. Appl Soft Comput 10(2):653–660
74. Dorigo M, Di Caro G (1999) Ant colony optimization: a new meta-heuristic. In: Proceedings of the 1999 congress on evolutionary computation-CEC99 (Cat. No. 99TH8406), vol 2. IEEE, pp 1470–1477. (1999, July)
75. Siahmarzkooh AT (2021) ACO-based type 2 diabetes detection using artificial neural networks. Indian J Forensic Med Toxicol 15(1)
76. Ojha VK, Abraham A, Snášel V (2014) ACO for continuous function optimization: a performance analysis. In: 2014 14th international conference on intelligent systems design and applications. IEEE, pp 145–150. (2014, November)
77. Akarsu E, Karahoca A (2011) Simultaneous feature selection and ant colony clustering. Procedia Comput Sci 3:1432–1438
78. Dorigo M, Stutzle T (2004) Ant colony optimization. MIT Press, Cambridge, MA
79. Parpinelli RS, Lopes HS, Freitas AA (2002) Data mining with an ant colony optimization algorithm. IEEE Trans Evol Comput 6(4):321–332
80. Peng L, Chen Y, Yang B, Chen Z (2005) A novel classification method based on data gravitation. In 2005 international conference on neural networks and brain, vol 2. IEEE, pp 667–672. (2005, October)
81. Ganji MF, Abadeh MS (2010) Using fuzzy ant colony optimization for diagnosis of diabetes disease. In: 2010 18th Iranian conference on electrical engineering. IEEE, pp 501–505. (2010, May)
82. Iyer A, Jeyalatha S, Sumbaly R (2015) Diagnosis of diabetes using classification mining techniques. arXiv:1502.03774
83. Sun Y, Dong W, Chen Y (2017) An improved routing algorithm based on ant colony optimization in wireless sensor networks. IEEE Commun Lett 21(6):1317–1320
84. Su Y, Bai Z, Xie D (2021) The optimizing resource allocation and task scheduling based on cloud computing and Ant Colony Optimization Algorithm. J Ambient Intell Humaniz Comput 1–9
85. Chen H, Tan G, Qian G, Chen R (2018, July). Ant colony optimization with tabu table to solve TSP problem. In: 2018 37th Chinese control conference (CCC). IEEE, pp 2523–2527. (2018, July)
86. Sagban R, Marhoon HA, Alubady R (2020) Hybrid bat-ant colony optimization algorithm for rule-based feature selection in health care. Int J Electric Comput Eng (IJECE) 10(6):6655–6663
87. Korb O, Stützle T, Exner TE (2006, September) PLANTS: application of ant colony optimization to structure-based drug design. In: International workshop on ant colony optimization and swarm intelligence. Springer, Berlin, Heidelberg, pp 247–258. (2006, September)
88. Sekhar SRM, Matt SG, Kumara Mahadevachar V (2023) Protein tertiary structure prediction by integrating ant colony optimization with path relinking and structure knowledge. Int J Inf Technol 15(3):1399–1405
89. Fidanova S, Lirkov I (2008, October). Ant colony system approach for protein folding. In: 2008 international multiconference on computer science and information technology. IEEE, pp 887–891. (2008, October)

90. Iqbal N, Kumar P (2023) From data science to bioscience: emerging era of bioinformatics applications, tools and challenges. Procedia Comput Sci 218:1516–1528
91. Wong KY (2008) Parameter tuning for ant colony optimization: a review. In: 2008 International conference on computer and communication engineering. IEEE, pp 542–545. (2008, May)
92. Aghdam MH, Ghasem-Aghaee N, Basiri ME (2009) Text feature selection using ant colony optimization. Expert Syst Appl 36(3):6843–6853
93. Elhoseny M, Shankar K, Uthayakumar J (2019) Intelligent diagnostic prediction and classification system for chronic kidney disease. Sci Rep 9(1):9583
94. Zhang XL, Chen XF, He ZJ (2010) Fault diagnosis based on support vector machines with parameter optimization by an ant colony algorithm. Proc Inst Mech Eng C J Mech Eng Sci 224(1):217–229
95. Li X, Zhang X, Li C, Zhang L (2013) Rolling element bearing fault detection using support vector machine with improved ant colony optimization. Measurement 46(8):2726–2734
96. Srilatha K, Ulagamuthalvi V (2020) Ant colony optimization based algorithm for detection of ovarian tumor. Int J Sci Technol Res 9(02):26422647

Chapter 5
Opposed Pheromone Ant Colony Optimization for Property Identification of Nonlinear Structures

Salar Farahmand-Tabar and Sina Shirgir

1 Introduction

Metaheuristic algorithms represent powerful optimization techniques utilized to address complex engineering problems that traditional optimization methods find challenging or even infeasible to solve [1–8]. These algorithms rely on diverse mathematical and computational concepts and are purposefully designed to efficiently explore the search space, identify promising solutions, and eventually converge toward the optimal solution. Nevertheless, while metaheuristic algorithms are effective, they do have certain limitations concerning efficiency, accuracy, and scalability. For instance, some of these algorithms might exhibit slow convergence or become stuck in local optima, while others may require a large number of fitness evaluations to attain the optimum solution. Furthermore, by the complexity increase in the problem, the performance of metaheuristic algorithms may deteriorate, rendering them less effective for real-world engineering challenges.

Hence, there exists a necessity to enhance the performance of metaheuristic algorithms to improve their efficiency and effectiveness when dealing with intricate engineering problems. This can be achieved through the development of innovative algorithms that address the limitations of existing algorithms or by integrating various techniques [9–16], such as machine learning, opposition-based learning, and adaptive parameter control, into the current algorithms to improve their performance. This endeavor to improve metaheuristic algorithms can have significant implications across diverse engineering fields, including mechanical, civil, and electrical engineering, where optimization plays a pivotal role in solving intricate problems. The

S. Farahmand-Tabar (✉)
Department of Civil Engineering, Faculty of Engineering, University of Zanjan, Zanjan, Iran
e-mail: farahmandsalar@znu.ac.ir; farahmandsalar@gmail.com

S. Shirgir
Department of Structural Engineering, Faculty of Civil Engineering, University of Tabriz, Tabriz, Iran

N. Dey (ed.), *Applications of Ant Colony Optimization and its Variants*, Springer Tracts in Nature-Inspired Computing, https://doi.org/10.1007/978-981-99-7227-2_5

outcomes of such improvements can lead to the creation of more efficient and reliable systems, cost reduction, and improvements in productivity and safety [17, 18].

Opposition-based learning (OBL) stands as a technique employed within optimization algorithms to generate opposite solutions to the existing population. In this approach, each individual solution in the population is paired with its opposite, generated by changing the sign of each decision variable's value. These opposite solutions are then evaluated and added to the population if they enhance the overall diversity or fitness. The underlying idea behind OBL is to diversify the population and prevent the optimization algorithm from becoming trapped in local optima. By generating opposite solutions, OBL enables more efficient exploration of the search space, leading to improved convergence and higher-quality solutions. Thus, OBL serves as a powerful technique that enhances the efficiency and effectiveness of optimization algorithms by promoting population diversity and facilitating more efficient search.

Numerous studies in the literature have successfully employed OBL to enhance various optimization algorithms, confirming its ability to boost algorithm performance [19]. For example, Sarkhel et al. incorporated OBL with the harmony search algorithm to accelerate its convergence speed [20]. Similarly, Shan et al. utilized OBL to improve both the convergence speed and the population diversity of the Bat Algorithm [21]. In a different context, Sapre and Mini employed OBL to enhance the convergence rate of the Moth Flame Optimization algorithm [22]. Additionally, Zhou et al. integrated OBL into the memetic algorithm to improve both its convergence speed and population diversity [23]. Chen et al. utilized OBL in the local search of the sine cosine approach for estimating parameters in photovoltaic models [24]. Raeesi et al. applied OBL to the salp swarm algorithm for the optimum design of control devices [25, 26]. Additionally, Shirgir et al. enhanced the searchability of charged system search algorithms using OBL in the identification of nonlinear systems [27].

The main perspective of this chapter is to improve the efficiency of the standard ACO algorithm that encounters problems such as being stuck in local optima due to a lack of population diversity. To boost the exploration and exploitation capabilities, opposition-based learning is incorporated into the standard versions of the algorithm, resulting in enhanced version. The proposed algorithm is then applied to identify the system of a nonlinear structure to find the mechanical properties such as mass and stiffness under seismic load. To assess the performance, a benchmark 8-story nonlinear frame is chosen, and the results of OPACO are compared across the basic ACO.

2 Related Works on System Identification

System identification poses a complex challenge in engineering encompassing parameter estimation [28], damage identification [29], and system behavior assessment [30, 31]. Structural system identification issues can be categorized into linear and nonlinear behaviors, leading to research addressing system identification and parameter estimation for various structural systems [32–36]. Nonlinear shear building structural systems are commonly encountered in engineering problems such as structural control [37, 38], behavior simplification [39], and parameter identification [40–42].

Dynamic systems with nonlinear behavior exhibit unique characteristics that are highly sensitive to instantaneous conditions, meaning that slight changes in system conditions can lead to drastic differences in future system behavior. As the state of nonlinear systems depends on momentary conditions, identifying their parameters requires exploring a large domain space. Consequently, the parameter identification problem for nonlinear systems must be framed within the context of large domain problems, necessitating a robust optimization algorithm.

3 Enhancing Method for Population Diversity

To improve optimization algorithms, especially concerning population diversity, a technique known as opposition-based learning (OBL) is employed. OBL utilizes opposition pairs to enhance the diversity of the population. This is achieved by generating a new individual as the mirror image of an existing individual with respect to the midpoint of the search space. By incorporating opposite characteristics, OBL effectively boosts population diversity, preventing the algorithm from being trapped in local optima and enhancing its ability to explore the solution space.

Integrating OBL with standard optimization algorithms leads to enhanced performance, resulting in faster convergence rates, better population diversity, and improved exploration and exploitation capabilities. These techniques are particularly valuable when tackling complex optimization problems that demand a delicate balance between exploration and exploitation, as commonly encountered in engineering design challenges.

3.1 Search Strategy of the Ant Colony

Ant colony optimization (ACO) is a nature-inspired metaheuristic that emulates the ants' foraging behavior to discover food sources [43]. ACO stands out among other metaheuristics due to its support for parallel processing, independence from process dependency, and valuable feedback it provides on the behaviors of ants within the search space [44].

When searching for food, ants are not visually impaired; ants have a remarkable ability to effectively locate the quickest path from their nest to a food source. During their movements, they release a chemical substance known as pheromone on the paths they traverse. This pheromone acts as a method of communication among ants, guiding them toward the shortest route leading to the food source. As ants travel, they can sense the presence of pheromone left by other ants, which leads them to favor paths with higher pheromone concentrations, thereby increasing the likelihood of more ants following the same path.

ACO utilizes two essential components, specifically, the heuristic data and pheromone trail, to arrive at probabilistic choices. As ants traverse a path, they update the pheromone level on each feature. Features with higher pheromone concentrations indicate a higher probability of belonging to the shortest path since more ants will traverse those paths. Consequently, the path exhibiting the highest level of pheromone becomes the most attractive and is chosen by a larger number of ants as the shortest path.

To initiate the process, each of the M features is initially assigned a pheromone value $\tau_0 = 1$, and a group of ants is randomly distributed among these features for a specified number of generations, denoted as T. At each generation g, the transition probability $TP_i^k(g)$ for the kth ant at the ith feature is determined following the guidelines provided in Refs. [45, 46].

$$TP_i^k(g) = \begin{cases} \frac{[\tau_i(g)]^\alpha . [\eta_i]^\beta}{\sum f_{j \notin j_i^k} [\tau_j(g)]^\alpha . [\eta_j]^\beta} & f_{j \notin j_i^k} \\ 0 & otherwise \end{cases} \quad (1)$$

The ants' exploration in the ant colony optimization (ACO) algorithm depends on the set of possible neighbors, j_i^k, which includes neighboring features that have not been visited by the kth ant. The ants' decision-making process is influenced by two non-negative parameters, α and β, which denote the relative importance of the pheromone level (τ_i) and heuristic information (η_j), respectively.

After an ant selects the subsequent feature in its path, the newly chosen set of features is assessed using a fitness function. If the fitness value does not improve upon adding any new feature, the movement of the kth ant is halted at that point. This process ensures that ants focus their efforts on more promising paths and avoid unproductive ones. However, if the termination criteria are not met, the level of pheromone at the ith feature for the next generation ($g + 1$) is updated according to the information provided in Ref. [43].

$$\tau_i(g+1) = (1-\rho)\tau_i(g) + \sum_{k=1}^{N} \Delta\tau_i^k(g)$$

$$\Delta\tau_i^k(g) = \begin{cases} \frac{fitness(S^k(g))}{|S^k(g)|}, & if\ i_\in S^k(g) \\ 0 & otherwise \end{cases} \quad (2)$$

The decay rate of pheromone ($0 \leq p \leq 1$), ants' number (N), the quantity $|S^k(g)|$ representing the number of chosen features, and $\Delta\tau_i^k$ signifying pheromone deposited by the kth ant is nonzero only when the ith feature is included in the ant's shortest path; otherwise, it remains at 0.

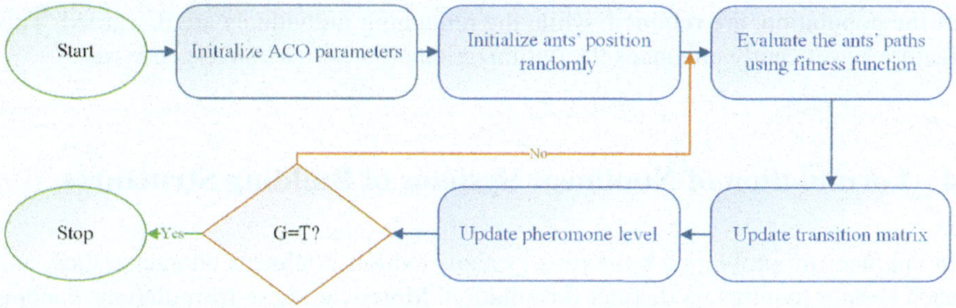

Fig. 1 Flow diagram of the ACO

The termination criteria are met when the generation count (g) reaches the predefined maximum value T. The optimal feature subset is determined by selecting the set of features with the highest level of pheromone and the smallest value of fitness. In other words, the combination of features that possesses both the most abundant pheromone trail and the best fitness evaluation is considered the optimum solution. The entire process of ant ACO is illustrated in Fig. 1.

3.2 *Opposed Pheromones in Ant Colony*

The opposition-based learning (OBL) method was proposed by Tizhoosh [19] within the realm of machine intelligence. This method revolves around creating an opposite counterpart of a given number, ensuring that it remains within specified upper and lower bounds. In simple terms, for a real number h within the range of a and b ($h \in [a, b]$), its opposite can be generated utilizing the equation as follows:

$$\overline{h} = b + a - h \tag{3}$$

Which can be generalized to n dimensions:

$$\overline{h}_i = b_i + a_i - h_i, i = 1, 2, \ldots, N \tag{4}$$

The opposite vector $\overline{h} \in R^n$ is obtained from the real vector $h \in R^n$. The fitness function is utilized to evaluate and compare the two solutions (h and $\overline{h}$), and the better solution is retained. For example, if the fitness function yields a better value for h compared to $\overline{h}$ (in the case of maximization), $\overline{h}$ is replaced with h; otherwise, h is kept.

Recently, there has been notable interest in the ability of OBL to accelerate the convergence rate of an optimizer. Specifically, OBL achieves this by doubling the population in each iteration, leading to an improved population. In every iteration, a new initial population is generated, and then the better individuals, comprising half of the population, are retained, while the remaining individuals are discarded. This feature significantly enhances the optimization algorithm's convergence rate.

4 Formulation of Nonlinear Systems of Building Structures

In engineering problems, numerous systems exhibit nonlinear characteristics, and each system requires its distinct formulation. Moreover, these formulations encompass specific parameters that govern the behavior of the system. One such illustration of nonlinear systems in engineering involves constructing structures with nonlinear behavior. Presented below is the formulation of such systems.

Within this section, the nonlinear dynamic analysis approach has been employed to address the equations governing dynamically excited structures. In the case of a nonlinear structure, the equation of motion is expressed as follows:

$$M\ddot{u}(t) + C\dot{u}(t) + F_S(t) = q(t) \tag{5}$$

Here, t represents time; $\ddot{u}$ and $\dot{u}$ are the acceleration and velocity vectors relative to the ground, respectively. M and C denote the mass and damping matrix, respectively, and F_S is the vector of resisting forces, which is a function of displacement. The term $q(t)$ represents the applied force, specifically in seismic cases, where it is defined by the ground acceleration time history.

Additionally, the initial conditions are given as $u(0) = u_0$ and $\dot{u}(0) = \dot{u}_0$, representing the initial displacement and velocity, respectively.

$$M\ddot{u}_{i+1} + C\dot{u}_{i+1} + F_{s\,i+1} = q_{i+1} \tag{6}$$

The analysis of nonlinear MDOF (Multiple Degrees of Freedom) structures has been conducted using the constant average acceleration method in combination with Newton–Raphson iteration. This approach is derived from time-stepping methods that consider the variation of acceleration over a time step [47].

The analysis begins by determining the initial state of the structural system (represented by K_{T0} and F_{s0}). Subsequently, the initial acceleration of the system is calculated as follows:

$$\ddot{u}_0 = M^{-1}\left(q_0 - C\dot{X}_0 - F_{s0}\right) \tag{7}$$

then for each time step:

$$\overline{q}_{i+1} = q_{i+1} + a_1 u_i + a_2 \dot{u}_i + a_3 \ddot{u}_i \cdot i = 0.1.2.\ldots \tag{8}$$

where:

$$a_1 = \left(\frac{1}{B(\Delta t)^2}\right)M + \left(\frac{G}{B\Delta t}\right)C \tag{9}$$

$$a_2 = \left(\frac{1}{B\Delta t}\right)M + \left(\frac{G}{B} - 1\right)C \tag{10}$$

$$a_3 = \left(\frac{1}{2B} - 1\right)M + \Delta t\left(\frac{G}{2B} - 1\right)C \tag{11}$$

When the resisting forces differ from the applied force, a residual force vector is established as:

$$\overline{R}_{i+1} = \overline{q}_{i+1} - (F_s)_{i+1} - a_1 u_{i+1} \tag{12}$$

The iterative method of Newton–Raphson is employed to calculate the extra displacement resulting from this residual force by solving:

$$\Delta u = \left(\overline{K}_T\right)^{-1}_{i+1} \times \overline{R}_{i+1} \tag{13}$$

The tangent stiffness, denoted as $\overline{K}_T$, can be represented as follows:

$$\left(\overline{K}_T\right)_{i+1} = (K_T)_{i+1} + a_1 \tag{14}$$

The following steps enable achieving the responses for a nonlinear structure:

$$u_{i+1} = u_i + \Delta u \tag{15}$$

$$\dot{u}_{i+1} = \left(\frac{G}{B\Delta t}\right)\Delta u + \left(1 - \frac{G}{B}\right)\dot{u}_i - \Delta t\left(1 - \frac{G}{2B}\right)u \tag{16}$$

$$\ddot{u}_{i+1} = \left(\frac{1}{B(\Delta t)^2}\right)\Delta u - \left(\frac{1}{B\Delta t}\right)\dot{u}_i + \left(1 - \frac{1}{2B}\right)\ddot{u}_i \tag{17}$$

where $G = 0.5$ and $B = 0.25$ are considered Newmark parameters for nonlinear analysis of the structure in this study.

5 Parameter Identification Problem

To precisely model the behavior of a system using a mathematical model, a system identification framework is employed to ascertain the model's unknown parameters. This involves minimizing the discrepancy between the actual response $f(t_i)$ of the system and the predicted response $\tilde{f}(t_i|\mathrm{p})$ of the model. To obtain this, an appropriate objective function is required for the optimization problem. The normalized mean square error (MSE) serves as a suitable objective function. Hence, for discrete-time t_i, the objective function based on MSE is expressed as follows [48]:

$$OF(\mathrm{p}) = \frac{\sum_{i=1}^{N}\left(f(t_i) - \tilde{f}(t_i|\mathrm{p})\right)^2}{N.\sigma_f^2} \tag{18}$$

Here, p denotes the unknown parameters of the system; σ_f^2 represents the variance of the actual response of the system, and N denotes the total number of data sets utilized in the identification process. It is important to emphasize that the identification process aims to minimize the objective function within the constraints defined for the parameter vector p.

$$p_{min} \leq p \leq p_{max} \tag{19}$$

where p_{min} and p_{max} denote the lower bound and upper bound of the system parameters, respectively.

6 Numerical Study: 8-Story Nonlinear Building

Opposition-based learning offers a means to explore a wider range of the search space, and employing simplexes such as Reflection, Expansion, Contraction, and Shrinkage enhances convergence during the extraction phase. To assess the algorithm's performance, a numerical study is needed, challenging both exploration and exploitation stages. The problem of identifying nonlinear systems with large domain falls under optimization problems that evaluate exploration by searching extensively and evaluate extraction using the MSE objective function with values approaching zero.

Numerous nonlinear models have been introduced and utilized in structural systems. The bilinear model is commonly used in structural analysis, while the Bouc-Wen model finds application in control devices. Hence, the objective is to identify the parameters of the bilinear model for a structural building system exhibiting nonlinear behavior based on seismic responses, which adds complexity to the problem. In this chapter, system parameters are identified for various domains, but the integration of these studies into optimization problems remains incomplete. However, the proposed methodology is applicable to other complex nonlinear models in system identification with a large domain search space.

Assuming knowledge of the seismic responses of the structure subjected to external excitation but with unknown mechanical parameters, optimization methods provide an efficient and computationally inexpensive way to solve system identification problems. An inverse scheme is used in this context, where seismic responses are first computed for specific mechanical property values, and then these numerical responses are treated as experimental data for system identification. In a structure, the mechanical parameters such as elastic stiffness, mass, damping, post-yield stiffness, and yield displacement, are treated as design variables to be determined such that the numerical responses align with the experimental responses. The objective function for this optimization problem is defined as the mean square error between the experimental and predicted responses.

The parameter identification problem for the nonlinear shear building structural system, in the form of an optimization problem, is stated as follows:

$$Find: Design\ Variables = \{m.k_1.k_2.u_y\ and\ c\}$$

$$To minimize: Objective\ Function = MSE = \frac{\sum_{i=1}^{N}\left(f(t_i) - \tilde{f}(t_i|\mathrm{p})\right)^2}{N.\sigma_f^2}$$

$$Subject\ to: \begin{cases} m_{min} \leq m \leq m_{max} \\ k_{2min} \leq k_1 \leq k_{1max} \\ k_{2min} \leq k_2 \leq k_{2max} \\ u_{y\,min} \leq u_y \leq u_{y\,max} \\ c_{d\,min} \leq c_d \leq c_{d\,max} \end{cases}$$

Numerically generated experimental data from seismic responses are utilized for system identification, serving as a test to assess the algorithm's search capability in finding specific parameters. To assess the proposed optimization algorithm's performance in solving identification problems for structural systems with nonlinear behavior, a shear frame structure model is considered, consisting of an 8-story structure with mass concentrated at each floor level. The model incorporates the inelastic behavior of materials through a bilinear force–displacement model, including elastic and post-yield stiffness, and yield displacement. Seismic analysis of the structural model is conducted using nonlinear time history analysis with the Newmark method.

Table 1 presents the mechanical properties of the structural model, such as stiffness, mass, and damping, for all stories. A damping ratio of 0.5% is utilized for the shear building. The mechanical parameters remain constant for all stories, and the post-yield stiffness is assumed to be 10% of the initial stiffness for each story, resulting in a post-yield stiffness of $K_p = 340.4 \times 10^2$ kN/m. Additionally, the yield displacement value is consistent at 2.4 cm for every story. A schematic image of the studied structural model and a bilinear model representing the nonlinear behavior of the structure are depicted in Fig. 2.

Table 1 Mechanical properties of 8-story nonlinear shear building [49]

Story	Mass (*ton*)	Stiffness (10^3 kN/m)	Damping (kN · s/m)
1–8	345.6	340.4	734.3

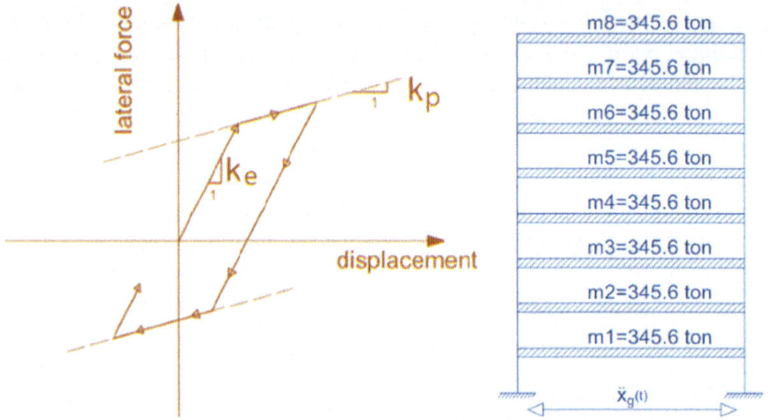

Fig. 2 Schematic representation of the structural model and bilinear behavioral model [49]

To gather numerically generated experimental data for the nonlinear seismic responses of the shear building, white noise ground acceleration (Fig. 3) with a PGA of 0.4 g, denoted as $W(t)$, is applied. Figure 4 displays the time history of the inter-story drift and displacement–force hysteresis cycle for the first story, which are used for parameter identification of the structural model.

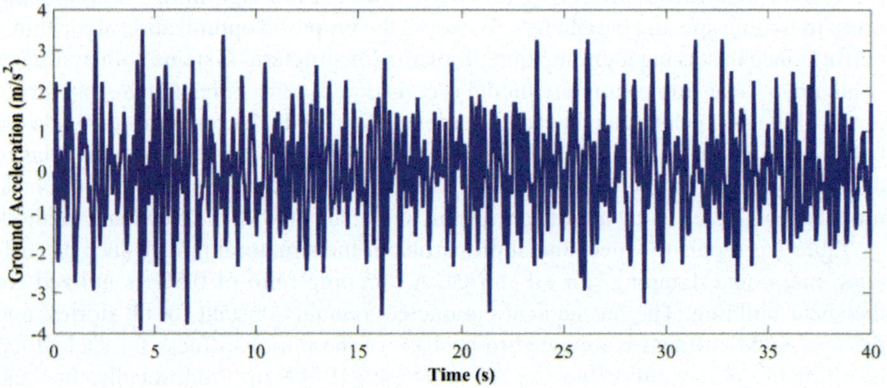

Fig. 3 The time history of ground acceleration (white noise), PGA = 0.4 g

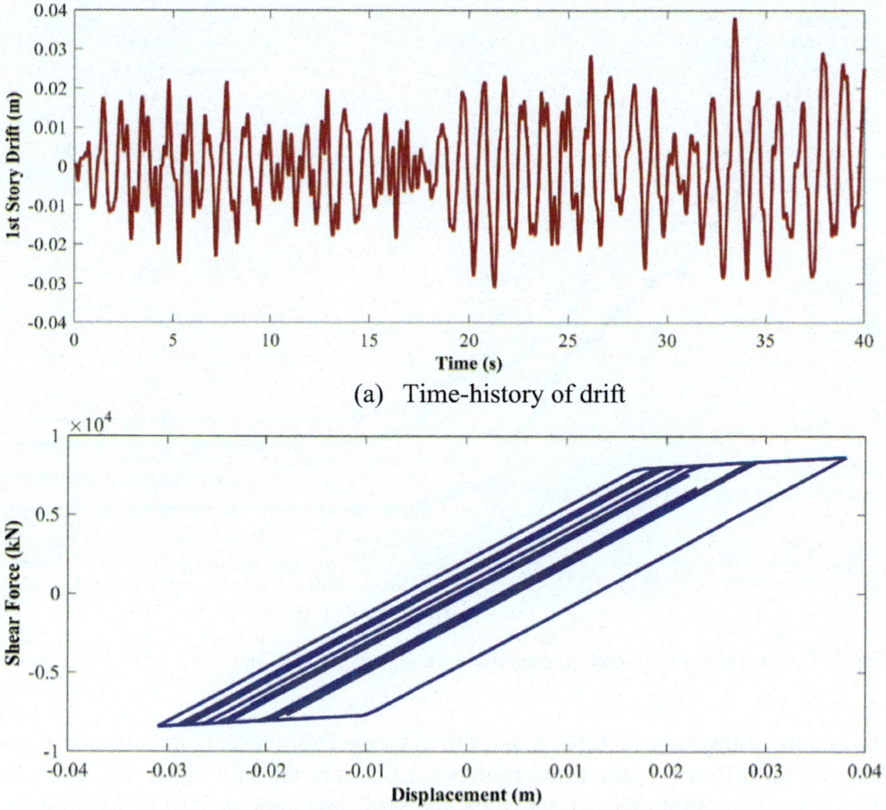

(b) Displacement–force hysteresis cycle

Fig. 4 Numerically generated experimental data of the 8-story shear building model under $W(t)$

7 Results and Discussion

In this section, the proposed OPACO algorithm and ACO are utilized to discover the optimal values for the mechanical parameters of the nonlinear shear building model under seismic excitation. The convergence history of the MSE for ACO and OPACO is illustrated in Fig. 5 on a logarithmic scale for different domains of variables.

The achieved values for these parameters and the results of the MSE criterion, as determined by the aforementioned optimization algorithms, are presented in Table 2. To comprehensively evaluate the algorithm's performance, a statistical test was conducted according to the best and mean solutions, the corresponding standard deviation, and the fewest iterations required for convergence. This statistical test was performed for all three domains of 50, 100, and 200% of the search space. Based on

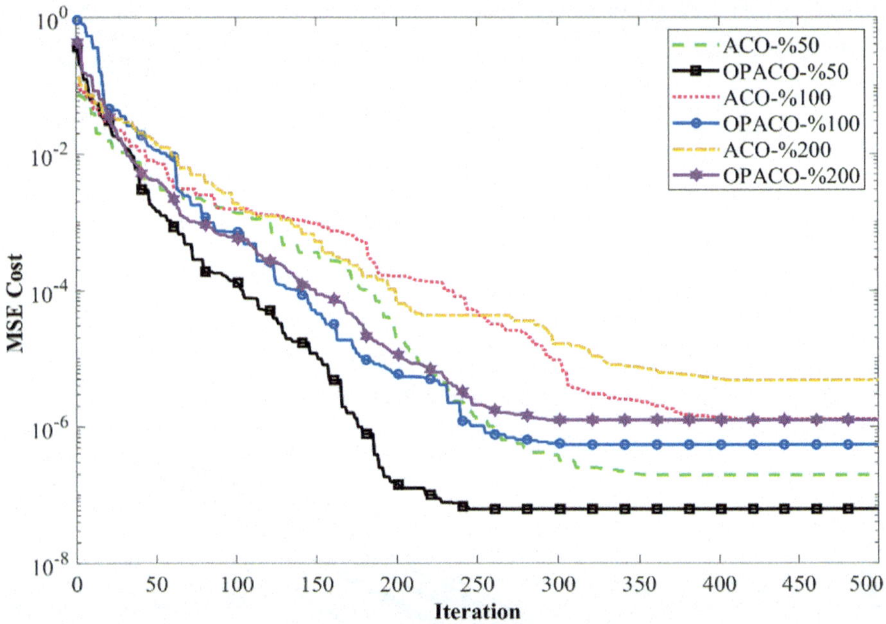

Fig. 5 Comparison of MSE convergence history based on domain limit

the results of this test, in Table 2, the OPACO algorithm reveals its robustness and stability for different cases. For a brief description, in the case of the 50% domain, the OPACO algorithm has a relative 68.4% MSE less than the ACO. Additionally, for the case of the 100% domain, the OPACO algorithm has a relative 58.3% MSE less than the ACO. Furthermore, for the 200% domain case, the OPACO algorithm

Table 2 MSE cost for various search domains using the ACO and OPACO

Domain limits (%)		ACO	OPACO
50	Best	1.9645E-07	6.2156E-08
	Mean	1.5468E-06	4.4984E-07
	STD	9.0152E-07	1.9715E-07
	C_{best}	357	245
100	Best	1.3016E-06	5.4287E-07
	Mean	2.4815E-05	4.1578E-06
	STD	1.0014E-05	2.3874E-06
	C_{best}	410	302
200	Best	4.8541E-06	1.2645E-06
	Mean	5.2549E-05	9.6841E-06
	STD	3.6471E-05	8.7245E-06
	C_{best}	404	293

has a relative 73.9% MSE that is less than that of ACO. According to the obtained results of the objective function, it is observed that the OPACO maintains its ability to solve large domain optimization problems.

As previously mentioned, engineering optimization problems are defined in various search domains to examine some features of the proposed OPACO. The results in Tables 3, 4 and 5 show this enhanced method's advantage in solving large domain optimization problems. Therefore, the results of system identification for a nonlinear shear building for a 200% search domain are presented in the following.

The time history of inter-story drift is compared in Fig. 6 for the experimental data and identified system using ACO and OPACO.

Additionally, the hysteresis bilinear force–displacement behavior is evaluated in Fig. 7 for the experimental data and identified systems. Finally, the errors for the predicted responses of the building model are assessed to show the accuracy of the identified system.

Table 3 Identified parameters set for nonlinear shear building with %50 domain

Parameter	Unit	Identification algorithm			
		ACO		OPACO	
		Value	Error%	Value	Error%
m	ton	359.93	4.148	349.71	1.189
k_1	10^3 kN/m	352.68	3.607	344.59	1.231
k_2	10^3 kN/m	35.402	4.002	34.471	1.265
u_y	cm	2.414	0.601	2.407	0.319
c	$kN \cdot s/m$	760.51	3.569	742.05	1.056
Average error %	–	–	3.185	–	1.012
MSE	–	1.9645E-07		6.2156E-08	

Table 4 Identified parameters set for nonlinear shear building with %100 domain

Parameter	Unit	Identification algorithm			
		ACO		OPACO	
		Value	Error%	Value	Error%
m	ton	312.82	9.485	332.51	3.788
k_1	10^3 kN/m	307.43	9.686	327.59	3.764
k_2	10^3 kN/m	38.869	14.187	35.96	5.631
u_y	cm	2.433	1.366	2.413	0.552
c	$kN \cdot s/m$	665.01	9.437	707.69	3.623
Average error %	–	–	8.832	–	3.472
MSE	–	1.3016E-06		5.4287E-07	

Table 5 Identified parameters set for nonlinear shear building with %200 domain

Parameter	Unit	Identification algorithm			
		ACO		OPACO	
		Value	Error%	Value	Error%
m	ton	398.52	15.312	371.90	7.610
k_1	10^3 kN/m	392.48	15.301	366.20	7.579
k_2	10^3 kN/m	39.237	15.269	36.557	7.395
u_y	cm	2.405	0.221	2.404	0.181
c	$kN \cdot s/m$	8.449	15.062	788.98	7.447
Average error %	–	–	12.233	–	6.042
MSE	–	4.8541E-06		1.2645E-06	

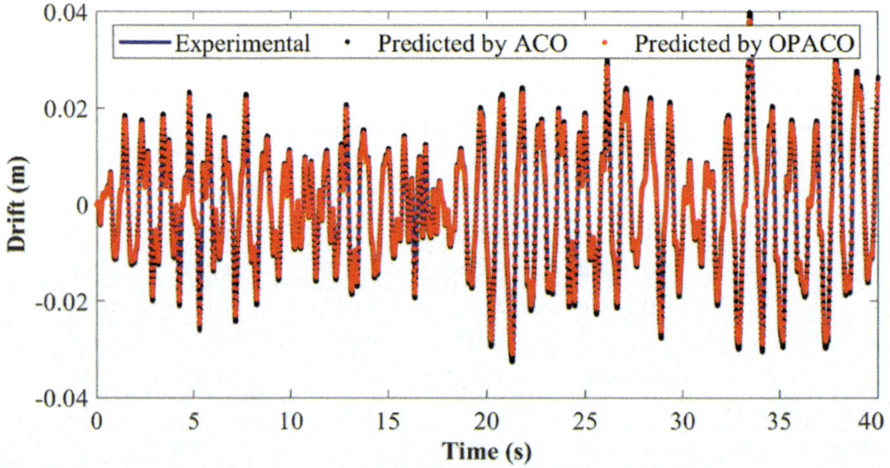

Fig. 6 Comparison of drift response of building with identified system for 200% domain

The time history of the error and the distribution of the error are depicted in Fig. 8. To display the OPACO performance, a Gaussian normal distribution was fitted to the error distributions. Figure 8 demonstrates that the system identified using OPACO with a standard deviation of 0.0058 is more appropriate than the system identified with ACO with a standard deviation of 0.0285 mm.

8 Conclusions

In this chapter, opposed pheromone ant colony optimization (OPACO) introduces an enhanced approach to ant colony optimization (ACO) through opposition-based learning and the concept of opposed pheromones. The algorithm combines the

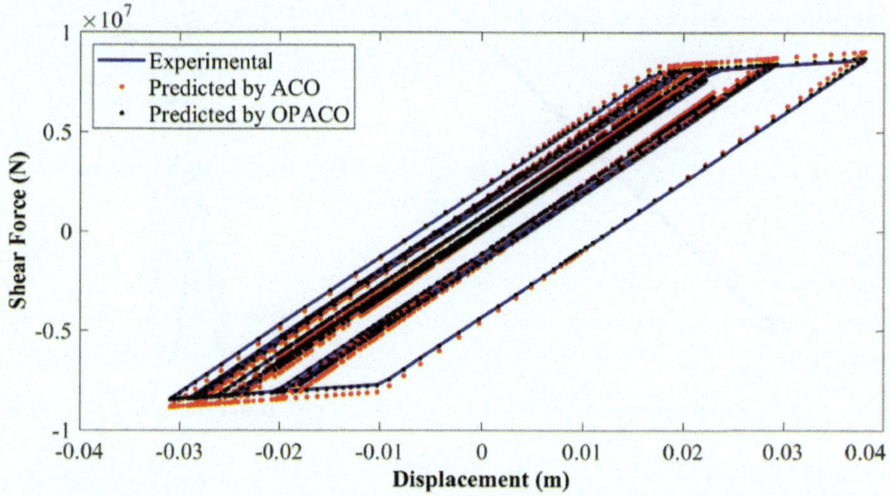

Fig. 7 Comparison of hysteresis bilinear force–displacement behavior for 200% domain

exploration and exploitation behavior of ants while introducing the idea of opposed pheromones to encourage diversification in the search space. This diversification leads to a more comprehensive exploration, enabling the algorithm to achieve improved convergence and the proper solution.

The effectiveness of OPACO is demonstrated in the context of system identification for nonlinear structures, specifically for finding mechanical properties such as mass and stiffness under seismic load. The algorithm utilizes the bilinear model of the utilized elements to represent the nonlinearity of the structure. The results clearly show that OPACO outperforms traditional ACO in terms of efficiency and accuracy in identifying solutions for structural optimization.

By leveraging opposition-based learning and incorporating the notion of opposed pheromones, OPACO offers a powerful and effective technique that significantly advances the field of structural optimization. Its ability to explore diverse regions of the search space and achieve superior solutions makes it a promising and valuable tool for various optimization problems beyond structural identification.

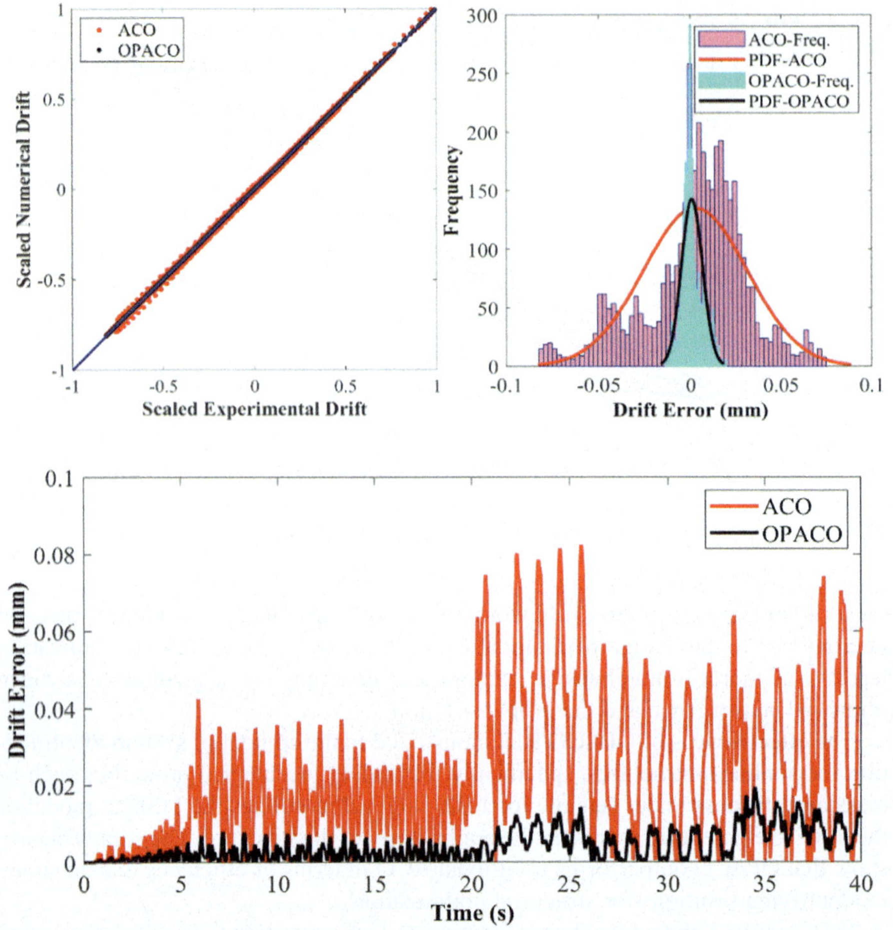

Fig. 8 Comparison of inter-story drift error for 200% domain

References

1. Farahmand-Tabar S, Ashtari P (2020) Simultaneous size and topology optimization of 3D outrigger-braced tall buildings with inclined belt truss using genetic algorithm. Struct Design Tall Spec Build 29(13):e1776. https://doi.org/10.1002/tal.1776
2. Ashtari P, Karami R, Farahmand-Tabar S (2021) Optimum geometrical pattern and design of real-size diagrid structures using accelerated fuzzy-genetic algorithm with bilinear membership function. Appl Soft Comput 110:107646. https://doi.org/10.1016/j.asoc.2021.107646
3. Farahmand-Tabar S (2023). Genetic algorithm and accelerating fuzzification for optimum sizing and topology design of real-size tall building systems. In: Dey N (ed) Applied genetic algorithm and its variants. Springer tracts in nature-inspired computing. Springer, Singapore. https://doi.org/10.1007/978-981-99-3428-7_9

4. Farahmand-Tabar S, Babaei M (2023) Memory-assisted adaptive multiverse optimizer and its application in structural shape and size optimization. Soft Comput. https://doi.org/10.1007/s00500-023-08349-9
5. Farahmand-Tabar S, Abdollahi F, Fatemi M (2023) Robust conjugate gradient methods for non-smooth convex optimization and image processing problems. In: Kulkarni AJ, Gandomi AH (eds) Handbook of formal optimization. Springer, Singapore. https://doi.org/10.1007/978-981-19-8851-6_42-1
6. Farahmand-Tabar S, Ashtari P (2023) Bilinear Fuzzy Genetic algorithm and its application on the optimum design of steel structures with semi-rigid connections. In: Kulkarni AJ, Gandomi AH (eds) Handbook of formal optimization. Springer, Singapore
7. Farahmand-Tabar S (2023) Memory-driven metaheuristics: improving optimization performance. In: Kulkarni AJ, Gandomi AH (eds) Handbook of formal optimization. Springer, Singapore. https://doi.org/10.1007/978-981-19-8851-6_38-1
8. Farahmand-Tabar S, Shirgir S (2023) Incorporating nelder mead simplex as an accelerating operator to improve the performance of metaheuristics in nonlinear system identification. In: Kulkarni AJ, Gandomi AH (eds) Handbook of formal optimization. Springer, Singapore. https://doi.org/10.1007/978-981-19-8851-6_39-1
9. Farahmand-Tabar S, Sadrekarimi N (2023) Overcoming constraints: the critical role of penalty functions as constraint handling methods in structural optimization. In: Kulkarni AJ, Gandomi AH (eds) Handbook of formal optimization. Springer, Singapore. https://doi.org/10.1007/978-981-19-8851-6_40-1
10. Shirgir S, Shamsaddinlou A, Zare RN, Zehtabiyan S, Bonab MH (2023) An efficient double-loop reliability-based optimization with metaheuristic algorithms to design soil nail walls under uncertain condition. Reliab Eng Syst Saf 232:109077
11. Shamsaddinlou A, Shirgir S, Hadidi A, Azar BF (2023) An efficient reliability-based design of TMD & MTMD in nonlinear structures under uncertainty. Structures 51:258–274
12. Farahmand-Tabar S, Rashid TA (2023) Steel plate fault detection using the fitness dependent optimizer and neural networks. In: Kulkarni AJ, Gandomi AH (eds) Handbook of formal optimization. Springer, Singapore. https://doi.org/10.1007/978-981-19-8851-6_41-1
13. Farahmand-Tabar S, Shirgir S (2023) Synergistic collaboration of motion-based metaheuristics for the strength prediction of cement-based mortar materials using TSK model. In: Kulkarni AJ, Gandomi AH (eds) Handbook of formal optimization. Springer, Singapore. https://doi.org/10.1007/978-981-19-8851-6_43-1
14. Farahmand-Tabar S, Shirgir S (2023) Positron-enabled atomic orbital search algorithm for improved reliability-based design optimization. In: Kulkarni AJ, Gandomi AH (eds) Handbook of formal optimization. Springer, Singapore. https://doi.org/10.1007/978-981-19-8851-6_44-1
15. Farahmand-Tabar S, Ashtari P, Babaei M (2023) Dynamic intelligence of self-organized map in the frequency-based optimum design of structures. In: Kulkarni AJ, Gandomi AH (eds) Handbook of formal optimization. Springer, Singapore. https://doi.org/10.1007/978-981-19-8851-6_45-1
16. Farahmand-Tabar S, Shirgir S (2023) Boosting the efficiency of metaheuristics through opposition-based learning in optimum locating of control systems in tall buildings. In: Kulkarni AJ, Gandomi AH (eds) Handbook of formal optimization. Springer, Singapore. https://doi.org/10.1007/978-981-19-8851-6_37-1
17. Dey N, Ashour A, Bhattacharyya S (2020) Applied nature-inspired computing: algorithms and case studies. Springer, Singapore
18. Dey N (2018) Advancements in applied metaheuristic computing. Engineering Science Reference, IGI Global, Hershey, PA, USA
19. Tizhoosh HR (2005) Opposition-based learning: a new scheme for machine intelligence. In: international conference on computational intelligence for modeling, control and automation and international conference on intelligent agents, web technologies and internet commerce (CIMCA-IAWTIC'06). IEEE, pp 695–701. https://doi.org/10.1109/CIMCA.2005.1631345
20. Sarkhel R, Chowdhury TM, Das M, Nasipuri M (2017) A novel harmony search algorithm embedded with metaheuristic opposition based learning. J Intell Fuzzy Syst 32:3189–3199. https://doi.org/10.3233/JIFS-169262

21. Shan X, Liu K, Sun PL (2016) Modified bat algorithm based on lévy flight and opposition based learning". Sci Program. https://doi.org/10.1155/2016/8031560
22. Sapre S, Mini S (2019) Opposition-based moth flame optimization with Cauchy mutation and evolutionary boundary constraint handling for global optimization. Soft Comput 23:6023–6041. https://doi.org/10.1007/s00500-018-3586-y
23. Zhou Y, Hao JK, Duval B (2017) Opposition-based memetic search for the maximum diversity problem. IEEE Trans Evol Comput 21:731–745. https://doi.org/10.1109/TEVC.2017.2674800
24. Chen H, Jiao S, Heidari AA, Wang M, Chen X, Zhao X (2019) An opposition-based sine cosine approach with local search for parameter estimation of photovoltaic models. Energy Convers Manag 195:927–942. https://doi.org/10.1016/j.enconman.2019.05.057
25. Raeesi F, Shirgir S, Azar BF, Veladi H, Ghaffarzadeh H (2020) Enhanced salp swarm algorithm based on opposition learning and merit function methods for optimum design of MTMD. Earthq Struct 18(6):719
26. Raeesi F, Veladi H, Azar BF, Shirgir S, Jafarpurian B (2023) Upgraded salp swarm algorithm for optimal design of semi-active MR dampers in buildings. Struct Eng Mech 86(2):197
27. Shirgir S, Azar BF, Hadidi A (2020) Opposition based charged system search for parameter identification problem in a simplified Bouc-Wen model. Earthq Struct 18(4):493
28. Niola V, Palli G, Strano S, Terzo M (2019) Nonlinear estimation of the Bouc-Wen model with parameter boundaries: application to seismic isolators. Comput Struct 222:1–9. https://doi.org/10.1016/j.compstruc.2019.06.006
29. Ho LV, Nguyen DH, Mousavi M, De Roeck G, Bui-Tien T, Gandomi AH et al (2021) A hybrid computational intelligence approach for structural damage detection using marine predator algorithm and feedforward neural networks. Comput Struct 252:106568. https://doi.org/10.1016/j.compstruc.2021.106568
30. Bao C, Hao H, Li Z-X, Zhu X (2009) Time-varying system identification using a newly improved HHT algorithm. Comput Struct 87:1611–1623. https://doi.org/10.1016/j.compstruc.2009.08.016
31. Luo J, Hong Y, Li C, Sun Y, Liu X, Yan Z (2022) Frequency identification based on power spectral density transmissibility under unknown colored noise excitation. Comput Struct 263:106741. https://doi.org/10.1016/j.compstruc.2022.106741
32. Zhang Z, Koh C, Duan W (2010) Uniformly sampled genetic algorithm with gradient search for structural identification–Part I: global search. Comput Struct 88:949–962. https://doi.org/10.1016/j.compstruc.2010.05.001
33. Feng D, Sun H, Feng MQ (2015) Simultaneous identification of bridge structural parameters and vehicle loads. Comput Struct 157:76–88. https://doi.org/10.1016/j.compstruc.2015.05.017
34. Ni Y-C, Zhang F-L (2019) Fast Bayesian frequency domain modal identification from seismic response data. Comput Struct 212:225–235. https://doi.org/10.1016/j.compstruc.2018.08.018
35. Bahar A, Pozo F, Acho L, Rodellar J, Barbat A (2010) Parameter identification of large-scale magnetorheological dampers in a benchmark building. Comput Struct 88:198–206. https://doi.org/10.1016/j.compstruc.2009.10.002
36. Shirgir S, Azar BF, Hadidi A (2020) Reliability based simplification of Bouc-Wen model and parameter identification using a new hybrid algorithm. Structures 27:297–308
37. Mohajer Rahbari N, Farahmand Azar B, Talatahari S, Safari H (2013) Semi-active direct control method for seismic alleviation of structures using MR dampers. Struct Control Health Monit 20:1021–1042. https://doi.org/10.1002/stc.1515
38. Yu T, Mu Z, Johnson EA (2023) Real-time neural network based semi-active model predictive control of structural vibrations. Comput Struct 275:106899. https://doi.org/10.1016/j.compstruc.2022.106899
39. Wilkinson S, Thambiratnam D (2001) Simplified procedure for seismic analysis of asymmetric buildings. Comput Struct 79:2833–2845. https://doi.org/10.1016/S0045-7949(01)00150-X
40. Oh BK, Glisic B, Park SW, Park HS (2020) Neural network-based seismic response prediction model for building structures using artificial earthquakes. J Sound Vib 468:115109. https://doi.org/10.1016/j.jsv.2019.115109

41. Xu C, Chase JG, Rodgers GW (2014) Physical parameter identification of nonlinear base-isolated buildings using seismic response data. Comput Struct 145:47–57. https://doi.org/10.1016/j.compstruc.2014.08.006
42. Jang T (2013) A method for simultaneous identification of the full nonlinear damping and the phase shift and amplitude of the external harmonic excitation in a forced nonlinear oscillator. Comput Struct 120:77–85. https://doi.org/10.1016/j.compstruc.2013.02.008
43. Dorigo M, Birattari M, Stutzle T (2006) Ant colony optimization. IEEE Comput Intell Mag 1:28–39
44. Wu Y, Gong M, Ma W, Wang S (2019) High-order graph matching based on ant colony optimization. Neurocomputing 328:97–104
45. Dorigo M, Stützle T (2019) Ant colony optimization: overview and recent advances. In: Handbook of metaheuristics. Springer, Cham, Switzerland, pp 311–351
46. Kanan HR, Faez K, Taheri SM (2007) Feature selection using ant colony optimization (ACO): a new method and comparative study in the application of face recognition system. In: Industrial conference on data mining. Springer, Berlin/Heidelberg, Germany
47. Newmark NM (1959) A method of computation for structural dynamics. J Eng Mech Div 85:67–94. https://doi.org/10.1061/JMCEA3.0000098
48. Charalampakis A, Dimou C (2010) Identification of Bouc-Wen hysteretic systems using particle swarm optimization. Comput Struct 88:1197–1205. https://doi.org/10.1016/j.compstruc.2010.06.009
49. Mohebbi M, Joghataie A (2012) Designing optimal tuned mass dampers for nonlinear frames by distributed genetic algorithms. Struct Design Tall Spec Build 21:57–76. https://doi.org/10.1002/tal.702

Chapter 6
Utilizing Ant Colony Optimization to Construct an S-Box Based on the 2D Logistic-Sine Coupled Map

Serap Şahinkaya and Deniz Ustun

1 Introduction

The motivation behind this research is rooted in the ever-evolving landscape of digital communication and the increasing demand for secure data transmission. In this context, cryptography algorithms play a pivotal role in ensuring data privacy. Specifically, symmetric key cryptography, which employs block ciphers, is of paramount importance for secure communication. The Substitution Box (S-Box) is a vital component of these ciphers, and its non-linearity and resistance to algebraic attacks are important for the security of the cryptography scheme. As modern cryptography systems require rigorous security measures, researchers continuously strive to explore innovative approaches and methods to enhance data transmission security in digital communication technologies.

To meet the strict security necessities of the modern cryptography systems, researchers continuously make an effort to develop or explore novel approaches and methods. These efforts are clearly shown that it is still crucial important the effective and powerful cryptography techniques are developed for the transmission of data in the digital communication technologies. Drawing inspiration from the foraging behavior and pheromone trails of ants, the Ant Colony Optimization (ACO) algorithm [18] emerges as a potent swarm intelligence-based optimization approach. The ability of these nature-inspired optimization methods to find near-optimal

S. Şahinkaya
Department of Natural and Mathematical Science, Tarsus University, Engineering Faculty, Mersin 33400, Türkiye
e-mail: serap@tarsus.edu.tr

D. Ustun (✉)
Department of Computer Engineering, Tarsus University, Engineering Faculty, Mersin 33400, Türkiye
e-mail: denizustun@tarsus.edu.tr

N. Dey (ed.), *Applications of Ant Colony Optimization and its Variants*, Springer Tracts in Nature-Inspired Computing, https://doi.org/10.1007/978-981-99-7227-2_6

solutions in the solution spaces of complex problems makes it an important candidate for optimizing the construction of S-Boxes.

This work deals with ACO application to S-Box design. The ACO is a powerful swarm intelligence-based optimization approach known for its ability to find near-optimal solutions for complex problems. By leveraging this nature-inspired optimization method, the researchers aim to optimize the construction of S-Boxes with enhanced security properties.

In this study, a variant of the ACO [33] algorithm adapted to continuous optimization is employed to optimize the parameters associated with 2D Logistic-Sine Coupled Map (2D-LSCM) [47]. The chaotic behavior map of 2D-LSCM is utilized for effective shuffling of S-Box components and permutation during encryption. The outcomes of the research demonstrate the successful implementation of the ACO algorithm for constructing an S-Box using 2D-LSCM for image encryption purposes. The resulting S-Box exhibits high non-linearity, resistance to algebraic attacks, and efficient diffusion of input changes to output changes, meeting the desired properties for an image encryption scheme.

Furthermore, the utilization of the 2D-LSCM for shuffling the S-Box components and in the permutation phase of encryption contributes significantly to the overall security of the encryption. Encrypted data demonstrates enhanced resistance against various attacks commonly used to decipher encrypted data, including algebraic attacks. This showcases the robustness of the constructed S-Box, which successfully withstands attempts to exploit vulnerabilities in cryptographic systems. Overall, the research presents an effective and powerful cryptography technique that leverages the ACO algorithm and the 2D-LSCM to enhance the security of data transmission in digital communication technologies, further advancing the field of secure communication.

2 Related Works

Lambic [26] made a significant early contribution to the design of strong cryptographic S-Boxes by introducing a pioneering technique to construct S-Box based on discrete maps. Building on this work, Tanyildizi et al. [36] later proposed their own method for generating chaotic S-Boxes by optimizing the parameters of one-dimensional maps. Additionally, in [12] authors presented a robust algorithm for S-Box design, which leveraged the chaotic scaled Zhongtang system as its foundation. These contributions have played a crucial role in advancing the field of chaotic S-Box design.

Enhancing the limitations of random computational methods, the artificial intelligence (AI) optimization approach capitalizes on the distinct traits of each AI optimization algorithm to conduct the search phase. By employing AI optimization operators, this approach can learn from past moves, represented by the herd of population agents, leading to superior solutions compared to random search methods. Furthermore, AI optimization approaches often integrate internal mechanisms to tackle the problem of being stuck in local optima. As a result, numerous applications of AI optimization for S-box design presented in the literature.

The literature provides evidence in support of integrating chaotic maps into AI optimization algorithms to improve the performance of S-boxes. Chen [13] applied a single population-based AI optimization algorithm that utilized simulated annealing with swapping that basen on a chaotic map. In [2] authors employed the integration of some maps with the Ant Colony Algorithm for S-box generation. In another investigation, in [38] authors explored the use of a 6-dimensional compound hyperchaotic map in conjunction with the ABC optimization approach. In [39] authors utilized the similar method.

3 S-Box

The S-Box plays a pivotal role as a fundamental element in contemporary symmetric key cryptography systems, particularly within the domain of block ciphers [3, 20, 42, 45]. By enabling a non-linear substitution operation, S-Box assumes a crucial role in facilitating the encryption and decryption of image. The principal objective of an S-Box is to induce chaos and augment the complexity of the encryption algorithm. By substituting input bit patterns with different output patterns, S-Boxes establish a highly non-linear relation among to the plaintext and ciphertext images, making it difficult for adversaries to deduce the original data or discern any patterns. High non-linearity is vital for achieving strong encryption. S-Boxes are typically constructed using various techniques, such as mathematical functions, lookup tables, or logical operations. Also, nature-inspired optimization algorithms and chaotic functions can be widely used in the construction of S-Box.

4 Two-Dimensional Logistic-Sine Coupled Map (2D-LSCM)

Chaotic maps have proven to be valuable tools in constructing S-Boxes, which are essential components in modern cryptography systems [47]. By utilizing chaotic maps, S-Boxes can achieve high levels of non-linearity, diffusion, and resistance against various cryptography attacks and chaotic maps, known for their sensitive dependence on initial parameters and unpredictable attitude, offer unique advantages in generating S-Boxes with desirable cryptography properties. When constructing S-Boxes using chaotic maps, it is crucial to carefully design the parameters and initial conditions to achieve desirable cryptography properties. The resulting S-Boxes should exhibit high non-linearity, resist linear and differential cryptanalysis, and possess strong confusion and diffusion properties. Additionally, the S-Boxes should be robust to attacks such as algebraic attacks, linear approximations, and differential attacks. Therefore, it is crucial important that use of artificial intelligence optimization algorithms in the optimization of parameters of chaotic maps to design S-Boxes with high levels of non-linearity, diffusion, and resistance against various cryptography attacks. The 2D-LSCM, introduced in 2018, exhibits remarkable properties such as high ergodicity, complex behavior, and an expansive chaotic range [47]. Due to its effective chaotic feature and classical confusion-diffusion structure ensuring a high level of security, powerful and secure S-Boxes can be constructed.

5 ACO Algorithm

ACO, drew inspiration from the food searching behavior by ants, is an algorithm rooted in swarm intelligence [18]. It was developed by Marco Dorigo in 1992 and has since gained popularity in solving various combinatorial optimization problems. ACO is particularly effective in solving problems that can be represented as graph-based and involve finding optimal paths or tours. The fundamental idea behind the ACO algorithm is to imitate the behavior of ants searching for food. Ants deposit a chemical substance known as pheromones as they traverse the ground, when they search for food. The pheromone is used as a tool that serves other ants to communicate with each other and the pheromone trail allows ants to determine the shortest route between the nest and food source. The ACO runs in the cycles and iterations as similar to other nature-inspired algorithms. During each iteration or cycle, virtual ants in the algorithm obtain their solutions by probabilistically choosing the next point to visit based on a combination of the pheromone level on the edges and a heuristic information about the quality of the path. Furthermore, the pheromone level on each edge is dynamically updated during each iteration or cycle, reflecting the solution quality discovered by ant. The possibility of a virtual ant selecting a particular point depends on two factors; the quantity of pheromone present on the edge connecting the present point to the future point and heuristic information that is generally the inverse of the distance between the points. In the algorithm, the updating rule of pheromone's amount is such that the edges of the better solutions are strengthened with more pheromone and over time, a positive feedback loop biasing the search toward better solutions is created. As the cycles or iterations progress, the pheromone trail on the edges evaporates, the algorithm actually simulates the consuming away of the pheromone over time. The ACO algorithm escapes from sub-optimal solutions and explores different regions of the search space, thanks to the evaporation process. The ACO continues its iterations until a finish condition is met, which could be a maximum iteration's number. During the search process, the best solution found by any virtual ant is considered the final solution to the optimization problem.

6 The Developed S-Box Construction

The steps including the stages related to the construction process are given in Fig. 1. In S-Box design, a 2D-LSCM which is known for its sensitive dependence on initial parameters and high complexity is utilized as the underlying chaotic system. In

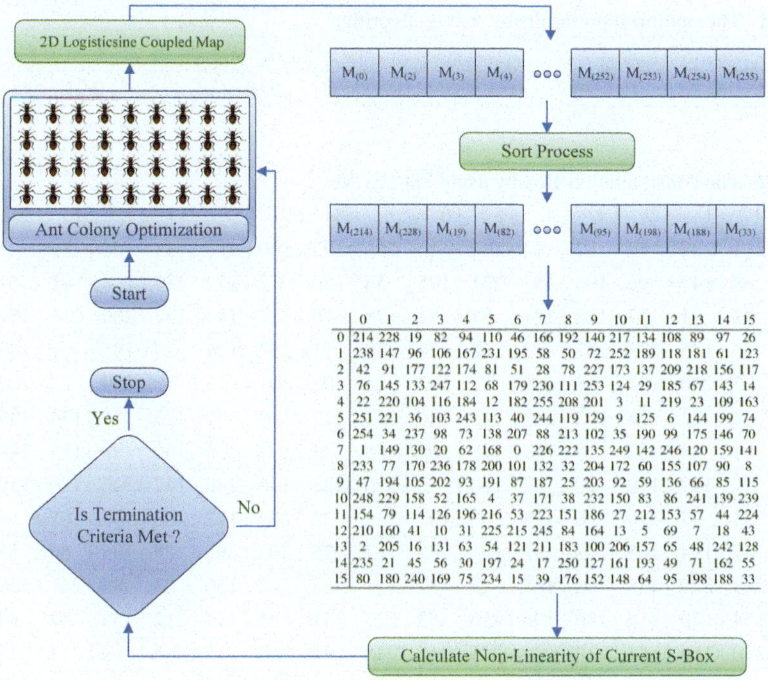

Fig. 1 A Construction flowchart of S-Box Based on 2D-LSCM by using ACO Algorithm

2023, the 2D chaotic map given in Eq. 1 with high complex behaviors is presented by Zhang et al. [47]. The control and adjustment parameters are optimized by using a variant of ACO ($ACO_{\mathbb{R}}$) algorithm adapted continuous problems [33]. By leveraging the principles of chaos theory and the collective intelligence of ant colonies, by employing this methodology, the resulting S-Box manifests favorable cryptographic features such as confusion, diffusion, and resistance to attacks.

$$\begin{aligned} x_{i+1} &= 10\beta\pi \left|cos\left[\pi\left(4ax_i\left(1-x_i\right)+\left(1-a\right)sin\left(\pi y_i\right)+b\right)\right]\right| \\ y_{i+1} &= 10\beta\pi \left|cos\left[\pi\left(4ay_i\left(1-y_i\right)+\left(1-a\right)sin\left(\pi x_{i+1}\right)+b\right)\right]\right| \end{aligned} \quad (1)$$

where $a\epsilon\,[0, 1]$ and $b\epsilon\,[0, 1]$ are control parameters of the 2D-LSCM map. The adjustment parameter β is equal to 10. However, in the optimization process, the value of β are chosen between 0 and 10 to obtain a more effective and flexible design structure in the construction of S-Box. The number of elements in the array generated by the map is equal to 256. The $ACO_{\mathbb{R}}$'s population is 30 and archive size k is equal to 60. The speed of convergence ξ is 1 and the locality of the search process q is chosen as 0.5. The iteration's number is equal to 100000. The optimization's results and constructed S-Box are shown in Tables 1 and 2, respectively.

Table 1 The optimization results by $ACO_{\mathbb{R}}$ algorithm

x_1	y_1	β	a	b
3.9158	102.9477	10	0.01	0.01

Table 2 The construction S-Box by using 2D-LSCM

	0	1	2	3	4	5	6	7	8	9	10	11	12	13	14	15
0	214	228	19	82	94	110	46	166	192	140	217	134	108	89	97	26
1	238	147	96	106	167	231	195	58	50	72	252	189	118	181	61	123
2	42	91	177	122	174	81	51	28	78	227	173	137	209	218	156	117
3	76	145	133	247	112	68	179	230	111	253	124	29	185	67	143	14
4	22	220	104	116	184	12	182	255	208	201	3	11	219	23	109	163
5	251	221	36	103	243	113	40	244	119	129	9	125	6	144	199	74
6	254	34	237	98	73	138	207	88	213	102	35	190	99	175	146	70
7	1	149	130	20	62	168	0	226	222	135	249	142	246	120	159	141
8	233	77	170	236	178	200	101	132	32	204	172	60	155	107	90	8
9	47	194	105	202	93	191	87	187	25	203	92	59	136	66	85	115
10	248	229	158	52	165	4	37	171	38	232	150	83	86	241	139	239
11	154	79	114	126	196	216	53	223	151	186	27	212	153	57	44	224
12	210	160	41	10	31	225	215	245	84	164	13	5	69	7	18	43
13	2	205	16	131	63	54	121	211	183	100	206	157	65	48	242	128
14	235	21	45	56	30	197	24	17	250	127	161	193	49	71	162	55
15	80	180	240	169	75	234	15	39	176	152	148	64	95	198	188	33

6.1 Performance Analysis for S-Box

The analysis and assessment of the S-Boxes are very crucial to prove their security and robustness against various attacks. To assess the efficacy of an S-Box within the encryption process, a range of performance metrics and criteria are employed, including non-linearity, linear approximation probability (LP), strict avalanche criterion (SAC), differential approximation probability (DAP), and bit independence criterion (BIC). These metrics and criteria serve as evaluative measures to gauge the S-Box effectiveness with regard to its cryptography properties and resistance against various attacks. The performance metrics are examined for the S-Box by the method.

6.1.1 Performance Analysis on Non Linearity

The primary characteristic of an S-Box that gauges its resilience against linear cryptanalysis is its non-linearity [11]. When an S-Box exhibits a high degree of non-linearity, even minor alterations in the input bits lead to substantial modifications in

Table 3 The non-linearity values of design S-Box and others

	1	2	3	4	5	6	7	8	Average
Developed S-Box	108	108	106	106	106	108	108	110	107.50
Ben Farah et al. [7]	106	106	106	108	106	104	106	108	106.25
Hoseini et al. [24]	104	98	106	104	104	104	104	106	103.75
Azkaynak [4]	106	106	104	102	106	108	106	104	105.25

the output bits. This heightened non-linearity enhances the overall security of the cryptographic algorithm employing this S-Box. If the average non-linearity value of an S-Box is greater than 106, it has enough strength to against cryptanalysis attacks. The evaluation of S-Box non-linearity can be performed through the Walsh spectrum (PM-Sosa, citation), which is represented as follows:

$$N(f) = 2^n(1 - 2^{-n} \max_{\omega \in GF(2^n)} \mid S_f(\omega) \mid),$$

here $S_f(\omega)$ indicates the Walsh-Hadamard transform of Boolean function f and it is computed as follows:

$$S_f(\omega) = \sum_{\omega \in GF(2^n)} (-1)^{f(x) \oplus x * \omega}.$$

The non-linearity values are tabulated in Table 3 for the S-Box by constructed our method and others. The non-linearity average value of the constructed S-box is equal to 107.25. The achieved average non-linearity value exceeds the non-linearity values of most S-Boxes constructed by other methods, as evidenced in Table 3. Hereby, the S-Box which is constructed by 2D-LSCM chaotic map and $ACO_{\mathbb{R}}$ illustrates a relatively high degree of non-linearity that is a desirable property for cryptography algorithms.

6.1.2 Performance Analysis on SAC

The SAC which is introduced by Webster and Tavares in 1986 [43] is a crucial measurement for evaluating the efficacy of an S-Box. The SAC measures the propagation of input bit changes to the output bits of an S-Box. In order to meet the SAC requirements, an S-Box should demonstrate a 50% likelihood of altering each output bit when a single input bit is modified. To put it another way, if we use the dependence matrix P to represent the SAC of an S-Box, the criterion is satisfied when each element P_{ij} of the matrix closely approximates 0.5 value. The S-Box meeting SAC criteria, establishes a strong basis for an effective encryption algorithm, providing a robust defense against various cryptography attacks. The offset value of SAC can be computed by the following formula:

Table 4 The tabulated results of the dependence matrix for design S-Box

0.5781	0.5469	0.5000	0.5000	0.4688	0.5000	0.4688	0.4844
0.4688	0.4375	0.5625	0.4688	0.5313	0.5938	0.5000	0.5625
0.5156	0.5156	0.4844	0.4844	0.4844	0.5156	0.4688	0.5156
0.5000	0.5313	0.5000	0.4844	0.4531	0.5156	0.5000	0.4688
0.4531	0.4688	0.4844	0.4688	0.4688	0.4531	0.5469	0.5469
0.5156	0.4688	0.4688	0.4688	0.4531	0.4531	0.5781	0.5781
0.5156	0.4844	0.5313	0.5469	0.5938	0.5313	0.4375	0.5000
0.5000	0.5000	0.5625	0.5938	0.4844	0.4844	0.5156	0.4844

$$\frac{1}{n^2} \sum_{1 \le i \le n} \sum_{1 \le j \le m} | 0.5 - P_{ij} | .$$

Table 4 illustrates the dependence matrix for the developed S-Box. The SAC value is calculated as 0.5039, and the generated S-Box exhibits an offset value of 0.126953.

6.1.3 Performance Analysis on BIC

BIC evaluates the statistical independence of input also output bits within S-Box. Essentially, it serves as a crucial metric for measuring independence between a specific pair of avalanche vectors and changes in input bits. In constructing S-Box, the eight employed functions contribute to establishing independent characteristics. The evaluation of BIC for S-Box involves computing the BIC-SAC value, which should be close to 0.5, and the BIC non-linearity value, which should be higher. BIC-SAC value closer to 0.5 indicates a stronger independence between the input and output bits, while a larger BIC non-linearity value signifies improved performance. BIC-SAC also BIC-non-linearity values for the developed S-Box are shown in Tables 5 and 6, respectively. Table 5 demonstrates that BIC-SAC values range from 0.4590 to 0.5273, with a mean value of 0.4995. On the other hand, the BIC non-linearity values range from 98 to 108, with a mean value of 103.99. The achieved values ensure evidence that the constructed S-Box by our method exhibits favorable performance in terms of the BIC criterion. The S-Box illustrates a high level of independence between input and output bits, enhancing its resistance against known-plaintext attacks and contributing to the overall security of the cryptographic algorithm.

Table 5 Tabulated results of BIC-SAC matrix values for developed S-Box

0	0.5137	0.5195	0.5039	0.4941	0.4980	0.5098	0.5039
0.5137	0	0.4648	0.5039	0.4980	0.4980	0.5273	0.5020
0.5195	0.4648	0	0.4902	0.5000	0.5176	0.4961	0.5000
0.5039	0.5039	0.4902	0	0.5176	0.5254	0.4922	0.4961
0.4941	0.4980	0.5000	0.5176	0	0.4590	0.4844	0.5020
0.4980	0.4980	0.5176	0.5254	0.4590	0	0.5000	0.5098
0.5098	0.5273	0.4961	0.4922	0.4844	0.5000	0	0.4590
0.5039	0.5020	0.5000	0.4961	0.5020	0.5098	0.4590	0

Table 6 Tabulated results of BIC-non-linearity matrix values for the developed S-Box

0	104	106	104	106	104	106	104
104	0	108	106	104	106	104	100
106	108	0	102	100	102	106	102
104	106	102	0	106	98	108	106
106	104	100	106	0	98	108	98
104	106	102	98	98	0	106	102
106	104	106	108	108	106	0	106
104	100	102	106	98	102	106	0

6.1.4 Performance Analysis on DAP

DAP measures S-Box's performance to withstand differential attacks. A decreased DAP value signifies a more robust S-Box with improved resistance against differential cryptanalysis. It means that the S-Box introduces significant complexity and non-linear transformations, making it more challenging for attackers to exploit differential characteristics and uncover sensitive information. By minimizing the DAP value, the image encryption algorithm achieves a higher level of security and confidentiality. Mathematically, differential uniformity of a map f is computed as follows:

$$DAP(f) = \max_{\Delta_x \neq 0, \Delta_y} \left(\frac{\#\{x \in GF(2^n) | f(x) \oplus f(x + \Delta_x) = \Delta_y\}}{2^n} \right),$$

here, Δ_x represents the input difference, and Δ_y represents the output difference. DAP matrix of the suggested S-Box can be seen in Table 7. Values in the developed S-Box range from 4 to 10, demonstrating its strong resilience against differential attacks. By achieving a minimum value of 4 and a maximum value of 10, the S-Box effectively thwarts attempts at differential cryptanalysis. This broad span signifies that even under various input differences, the S-Box consistently exhibits significant and unpredictable transformations in the output, fortifying its resistance against potential attacks.

Table 7 Tabulated results of the DAP matrix values for design S-Box

6	8	6	6	8	6	6	6	6	8	6	8	6	6	6	6
8	8	8	6	8	8	8	8	6	6	6	6	8	6	6	6
8	6	6	6	8	6	8	6	6	8	6	6	8	6	6	6
6	6	6	8	6	6	6	6	6	8	6	6	6	8	6	8
6	6	6	8	8	6	6	6	6	6	6	6	6	4	6	8
6	8	6	6	6	8	6	6	6	6	8	6	6	8	6	6
6	6	6	4	6	6	6	6	6	6	6	6	8	6	6	8
8	6	4	6	8	6	6	8	8	6	6	6	6	6	8	6
6	6	6	8	8	6	8	6	8	6	4	8	6	6	8	6
8	6	6	6	6	6	4	6	6	4	6	6	8	8	6	6
8	6	6	6	6	6	6	6	6	6	6	8	8	8	6	6
6	6	6	6	6	8	6	6	6	6	6	6	8	6	8	8
8	6	8	8	6	8	6	6	8	6	10	6	6	6	6	8
10	8	6	8	6	6	8	6	10	6	6	8	6	8	6	6
6	8	6	6	6	6	6	6	8	6	8	6	6	6	6	8
6	6	6	6	6	6	6	6	6	6	10	6	6	6	8	0

6.1.5 Performance Analysis on LP

LP evaluates the probability of linear characteristics occurring in an S-Box. It measures the likelihood of linear approximations being satisfied, indicating the susceptibility of our S-Box to linear cryptanalysis. A lower LP value indicates a higher resistance against linear attacks, thereby enhancing the overall security of the cryptography algorithm. The LP can be computed by the following mathematical expression:

$$LP = \max_{\Gamma_x \Gamma_y \neq 0} \left| \frac{\#\{x/x \cdot \Gamma_x = S(x) \cdot \Gamma_y = \Delta_y\}}{2^n} - \frac{1}{2} \right|,$$

Within this equation, the variable x corresponds to the collection encompassing all conceivable input values, whereas 2^n represents the overall count of elements contained within the set. The symbols Γ_x and Γ_y refer to the masks that operate on parity of input and output bits. By evaluating this formula for the developed S-Box, we determined that the maximum value of LP is 0.148438. This indicates the maximum deviation from the ideal value of $\frac{1}{2}$. The LP value serves as a measure of S-Box's susceptibility to linear attacks. A lower LP value suggests a higher resistance against such attacks, making the design S-Box a favorable choice for ensuring the security and integrity of cryptographic algorithms.

Table 8 The tabulated results of comparisons for the S-Boxes generated by different methods

	Non-linearity			Bit Independence Criteria (BIC)		SAC	DAP	FixPoint	Rev FixPoint	Iteration Ring
S-Box	min	max	avg	Non-linearity	SAC					
developed S-Box	106	110	**107.50**	103.93	0.4995	0.5039	10	0	0	10
BenFarah et al. [7]	104	108	106.25	103.64	0.4996	0.5009	10	–	–	–
Hoseini et al. [24]	98	106	103.75	103.80	0.4628	0.4965	–	–	–	–
Azkaynak [4]	102	108	105.25	102.64	0.4994	0.5037	10	–	–	–

6.1.6 Remarks on Performance Analyses for Design S-Box

Table 8 presents a comparison of different S-Boxes generated by various methods based on several evaluation criteria. The S-Boxes are evaluated in terms of non-linearity, BIC, SAC, DAP, FixPoint, Reverse FixPoint, and Iteration Ring [4, 7, 24]. Moreover, it is evident that the design S-Box exhibits promising performance. In terms of non-linearity, the design S-Box achieves an average value of 107.50, which is higher than the other methods, indicating stronger non-linearity. Additionally, it demonstrates competitive values for BIC and SAC, with scores of 0.4995 and 0.5039, respectively. Comparatively, the S-Boxes presented in [7, 24], and [4] also showcase respectable results. However, their non-linearity averages of 106.25, 103.75, and 105.25, respectively, fall slightly behind the design S-Box. In summary, the design S-Box outperforms the alternatives in terms of non-linearity, exhibiting higher average values. Although the other S-Boxes present notable results, the lack of provided data for certain criteria makes it difficult to make a conclusive comparison. Therefore, the developed S-Box demonstrates its potential to be a robust choice for cryptographic applications.

7 Utilization of the Created S-Box for Encryption

This section elucidates the application of design S-Box in encryption, providing a comprehensive explanation of the encryption outlined in [19]. The image encryption scheme consists of four steps: main key generation using secret and public keys, the Bit-Reversion and permutation steps based on 2D-LSCM and the S-Box-based diffusion step. The diffusion step based on S-Box remains unchanged from [19].

7.1 Production of Main Key

By combining the hash values of the time signature (h, m, s) using SHA-512 and MD5, we create a secret key. In a similar manner, the public key is derived from the plaintext image by utilizing a blend of SHA-512 and MD5 hash values. This process entails computing three vectors for the plaintext image: the row vector, the column vector, and the diagonal vector. The row vector $r \in R_{1\times 512}$ signifies the summation of pixel values along all rows. The column vector $c \in R_{512\times 1}$ signifies the summation of pixel values along all columns. The diagonal vector $d \in R_{1\times 1023}$ represents the sum of a diagonal in the image. Table 9 employs the notation (r, c^T, d) to indicate the row matrix derived from these computations, where c^T represents the transpose matrix. The primary key is obtained through the utilization of an XOR operation performed on the public and secret keys. Nevertheless, the main key itself is not directly utilized for encryption purposes. Instead, we derive a row matrix G with dimensions 1×4 by the main key, which is then employed in the diffusion phase and also permutation phase to carry out the encryption process. The procedure for computing the row matrix G is described in detail in Table 9.

7.2 Bit-Reversion Step

The implementation of a novel process called "bit reversion" given in Table 10 has been presented to significantly enhance the encryption security. The bit reversion operation involves applying an XOR among the diffused pixels and a matrix Y, which is obtained by reversing the bits of a matrix RC_s derived from a chaotic sequence. By incorporating this step, the encryption process becomes even more

Table 9 Generation of G by using main key

1: Compute the vector (r, c, d) by plaintext image
2: Compute a binary public key p_k of length 512 from the vector (r, c^T, d) by employing a combination of SHA-512 and MD5
3: Compute binary secret key s_k of length 512 from the time signature (h, m, s) by employing a combination of SHA-512 and MD5
4: Compute main key m_k by applying xor operation on p_k and s_k, that is, $m_k = xor(p_k, s_k)$
5: Reshape m_k to 4 matrices of size 8×16
6: For each 2×2 matrices compute sum of values of all columns in modulo 2
7: Reshape these 8 column matrices of size 4×1 to a matrix F of size 8×4
8: Compute 1×4 matrix G by the following matrix multiplication: G=(128 64 32 16 8 4 2 1) $F_{8\times 4}$

Table 10 Bit-Reversion steps

1: Compute chaotic sequence $C_s = [c_{1r}]_{1\times mn}$ by using x_0, y_0 and a, b
2: Calculate the row matrix $RC_s = [rc_{1r}]_{1\times mn}$ defined by $RC_s = floor(256 \times C_s)$
3: for all r, re-write $\tilde{rc}_{1r}$ in reverse order and set to $\tilde{y}_{1r}$ After, $Y = [y_{1r}]_{1\times mn}$ to $m \times n$ in shape
4: Calculate Bit Revise matrix $Br = [br_{sd}]_{m\times n}$ defined by $Br = bitxor(ImgMat, Y)$

robust and resistant to potential attacks. The resultant matrix Br, generated from the bit reversion process, represents the final ciphertext image, ensuring the utmost protection of sensitive data and cementing the study's contributions to advancing the field of encryption methodologies.

7.3 2D-LSCM-Based Permutation Phase

The permutation phase that based on 2D-LSCM is a highly efficient process designed to effectively alter the positions of pixel within an image, contributing to the overall strength and complexity of the encryption scheme. The detailed permutation step is illustrated in Table 11.

8 Analysis of Results and Security

To assess the effectiveness of the encryption technique utilizing the design S-Box, a series of experiments were conducted. These experiments involved the use of a 512 × 512 image called "4.2.07," obtained from the website mentioned in reference [37]. The encryption process was carried out meticulously and the encrypted images were subjected to comprehensive evaluation. Both the encryption and testing processes are

Table 11 Permutation steps

1: Take the row matrix G that is computed in Table 9
2: Calculate initial values of $x_0 = [G(1)/256]100\pi$, $y_0 = [G(2)/256]100\pi$
3: Calculate initial parameters of $a = G(3)/256$, $b = G(4)/256$
4: Calculate 2D-LSCM value for parameters (x_0, y_0, a, b)
5: Generate $X_1 \times (256 \times 256)$ by 2D-LSCM
6: Shape $Y_1 \times (256 \times 256)$ by ordering X in an ascending order
7: Reshape K matrix of size 256×256 as an K' matrix of size $1 \times (256 \times 256)$
8: Shape K' with respect to Y
9: Reshape K' as a 256×256 which is a permuted image matrix

executed on a high-performance workstation, featuring an Intel Xeon processor with a clock speed of 4.0 GHz and a substantial 64 GByte RAM. The encryption algorithm and performance analyses were employed by using the MATLAB software.

8.1 Results

The section represents results for image of 4.2.07.jpg, which has a resolution of 512 × 512 pixels, as shown in Fig. 2. Figure 2 clearly illustrates that the ciphered images resulting from the encryption process do not contain any discernible or meaningful information. This outcome demonstrates the successful implementation of the encryption method in concealing the content of the original image. Furthermore, the original image was recovered with high fidelity through the application of the corresponding decryption processes.

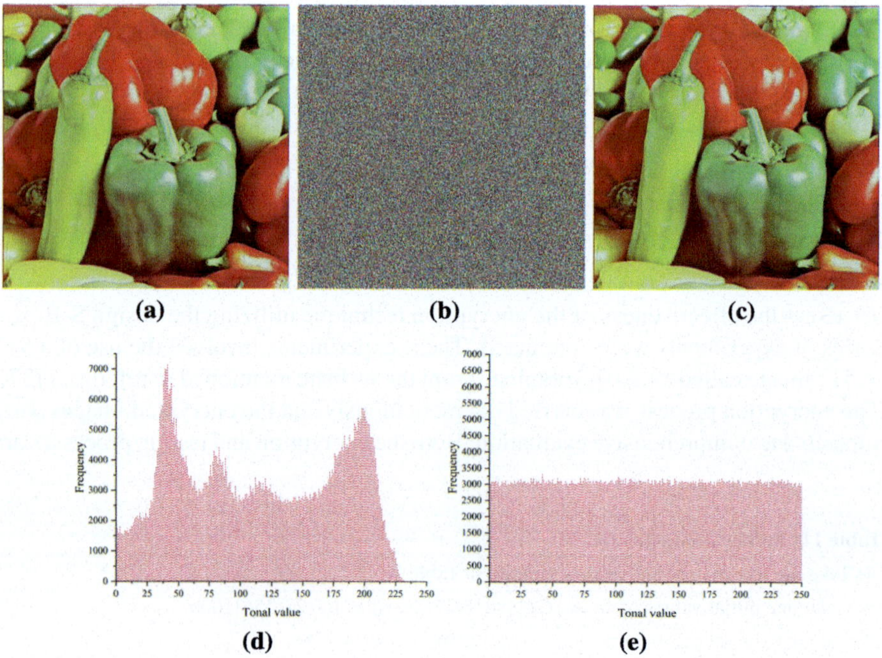

Fig. 2 Results for encryption: **(a)** Orginal **(b)** Chiper **(c)** Decrypted **(d)** Histogram of image in (a) **(e)** Histogram of image in (b)

8.2 Security Analysis

In the cryptography, the various cryptanalysis methods have been developed to analyze and crack cryptographic algorithms. Among the methods, entropy, histogram, correlation, key space and key sensitivity analyses, differential, noise, and cropping attacks are the most common and reliable methods. In this section, an exhaustive examination of these cryptanalysis techniques ensuring valuable insights into their effectiveness and applicability is carried out. Furthermore, the achieved results are meticulously compared with the presented in recent research papers, such as [6, 15, 46].

8.2.1 Exploring Image Distribution: Histogram Analysis

A histogram is a visual representation that provides an overview of the distribution of pixel values within an image. It ensures valuable insights into the overall characteristics of an image's pixel intensity levels. In the context of encryption, a histogram distribution in the cipher image which is uniform enhances the security against cyber attacks. When histograms of the 4.2.07 plaintext image given in Fig. 3, as well as their corresponding encrypted image, are evaluated, it appears that the cipher image exhibits a distribution that is uniform. This uniformity in the histogram further strengthens the security of the encryption process and safeguards against potential vulnerabilities in statistical analysis.

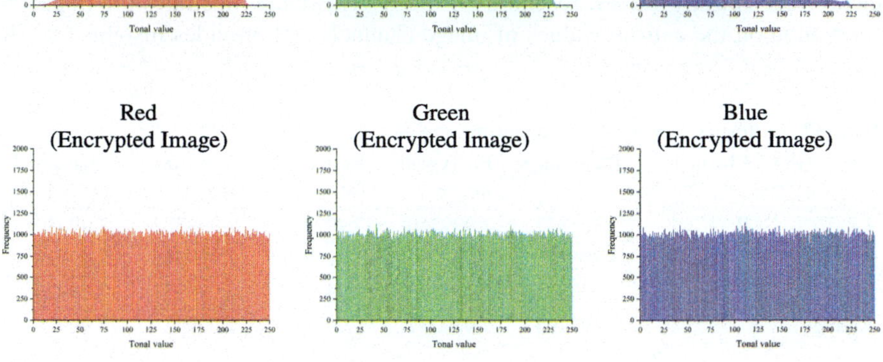

Fig. 3 The Graph of plaintext and ciphertext images of 4.2.07

8.2.2 Information Entropy Analysis

The measurement of entropies is a fundamental technique used to assess the uncertainties associated with ciphered images [30]. In evaluating the quality of an encryption algorithm, it is desirable to obtain an entropy value close to 8. This signifies efficient encryption, where the entropy of a given sequence s can be evaluated by the following formula:

$$H(s) = \sum_{i=1}^{l-1} P(s_i) \log_2 \frac{1}{P(s_i)},$$

where, where $l = 2^m$, and m represents the total samples, s_i denotes image's gray value. Here, variable $P(s_i)$ signifies the probability of s_i within the sequence s, while the summation $\sum_{i=1}^{l-1} P(s_i) = 1$ ensures that the probabilities of all elements in the sequence sum up to unity. The outcomes obtained from the entropy analysis are conveniently summarized in Table 12. This table likely presents the calculated entropy values associated with the image, reflecting the level of uncertainty/information content within the data. The entropy values provide valuable insights into the distribution and diversity of gray values present in the image, aiding in various image processing and analysis tasks.

Table 12 provides a comparison of information entropy analysis for various image channels, including plain and encrypted images, based on the findings of different studies [6, 15, 46]. The table presents the entropy values for RGB (color channels: red, green, and blue) of an image labeled 4.2.07 with dimensions 512×512. For the plain image, the entropy values are 7.338827 (R), 7.496253 (G), and 7.058305 (B). After encryption, the entropy values increase significantly to 7.999250 (R), 7.999286 (G), and 7.999175 (B). The study by [6] reports entropy values of 7.9972 (R), 7.9969 (G), and 7.9973 (B) for the encrypted image. Similarly, [15] and [46] present entropy values of 7.999382 (R), 7.999348 (G), and 7.999251 (B), and 7.99930 (R), 7.99929 (G), and 7.99920 (B), respectively. Overall, the average entropy values for the plain image is 7.297795, there is a noticeable and significant increase in the average entropy values, reaching 7.999237. This table illustrates the impact of encryption on the entropy values of image channels and provides insights into the

Table 12 Comparison of information entropy analysis

Image/H(s)	Channel	Plain image	Encrypted image	Basha et al. [6]	Demirtas [15]	Zhang and Gong [46]
4.2.07	R	7.338827	7.999250	7.9972	7.999382	7.99930
(512×512)	G	7.496253	7.999286	7.9969	7.999348	7.99929
	B	7.058305	7.999175	7.9973	7.999251	7.99920
Average		7.297795	7.999237	7.9971	7.9994	7.9993

Table 13 NPCR and UACI analysis

Image/H(s)	Channels	Pixel position	Changed pixel value	Our method
NPCR	R G B	(100, 475, 2)	200− > 201	99.61968% 99.60594% 99.60175% 99.61% (Avg.)
UACI	R G B	(100, 475, 2)	200− > 201	33.48271% 33.40381% 33.40878% 33.43% (Avg.)

effectiveness of different encryption techniques in preserving information entropy (Table 13).

8.2.3 Resistance Against Differential Attack

The differential attack is widely employed in cryptography to assess the encryption algorithms security. The effectiveness of encryption against cyber attack depends on its sensitivity to modifications in the original image. A minor alteration in the value of one or multiple pixels in the original image has the potential to entirely transform it into a distinct encrypted image. To evaluate an encryption algorithm's resistance to the differential attack, two commonly used requirements are NPCR and UACI as mentioned in [40]. When comparing two dissimilar images, the expected NPCR value is 99.6094.

8.2.4 Analysis of Pixel Correlation

The formula provided below is utilized to evaluate the coefficient of correlation between two adjacent pixels in an image:

$$r_{\alpha\beta} = \frac{cov(\alpha, \beta)}{\sqrt{\psi(\alpha)}\sqrt{\psi(\beta)}},$$

where

$$\psi(\alpha) = \frac{1}{N}\sum_{i=1}^{N}[\alpha_i - E(\alpha)]^2,$$

$$cov(\alpha, \beta) = \frac{1}{N}\sum_{i=1}^{N}[\alpha_i - E(\alpha)][\beta_i - E(\beta)].$$

In this context, the symbols α_i and β_i represent two neighboring pixels, while N represents the overall count of pixel pairs (α_i, β_i) obtained from the image. The terms $E(\alpha)$ and $E(\beta)$ correspond to the mean values of α_i and β_i respectively. Correlation analysis is commonly used to evaluate the relation among variables or image's pixels. In the context of image encryption, it has been observed that neighboring pixels in an unencrypted image demonstrate a significant correlation in three directions: horizontal, vertical, and diagonal. However, to enhance the resistance of an image encryption algorithm, it is essential to reduce the correlation among these neighboring pixels. To illustrate this phenomenon, Table 14 presents the correlation values between adjacent pixels in both the original unencrypted images and their corresponding encrypted versions. Additionally, Fig. 4 visually compares the pixel distributions of neighboring pixels along the three directions (horizontal, vertical, and diagonal) in the original and encrypted images. These measurements and visual representations offer valuable insights into the impact of encryption on the correlation and pixel distribution characteristics of the image data.

8.2.5 Cropping Attacks

A cropping attack is a malicious technique utilized by attackers with the intention of compromising the integrity of encrypted ciphertext data. The main goal of this attack is to impede the intended recipient from successfully decrypting the ciphertext. To ensure the efficiency and robustness of an image encryption algorithm, it is vital to minimize any potential corruption resulting from cropping attacks.

Figure 5 illustrates the simulation of various degrees of cropping attacks. The experiment involves decrypting the cropped version of the image "4.2.07.jpg" with dimensions 512×512. The image is subjected to cropping ratios of $1/2$, $1/4$, and $1/8$, resulting in partial removal of its content. Despite these cropping attacks, the decryption process manages to produce identifiable decrypted images, as depicted in Fig. 5.

In addition, Table 15 presents the PSNR scores of the decrypted images, which serve as an indicator of image quality. Higher PSNR values in the table correspond to lower levels of corruption. The calculation of the PSNR value is performed according to the formula specified in references [25, 41].

$$PSNR := 10\log(\frac{255^2}{MSE}),$$

within this context, the mean-squared error (MSE) is employed as a metric, and its calculation is based on the following formula:

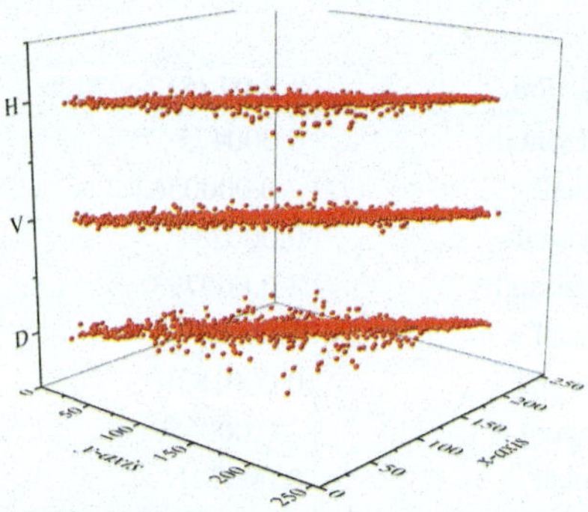

(a) Pixel distribution of plaintext image for red channel

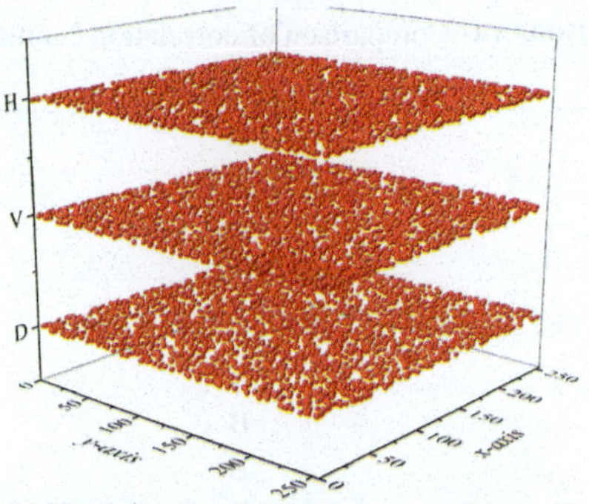

(b) Pixel distribution of cipher text image for red channel

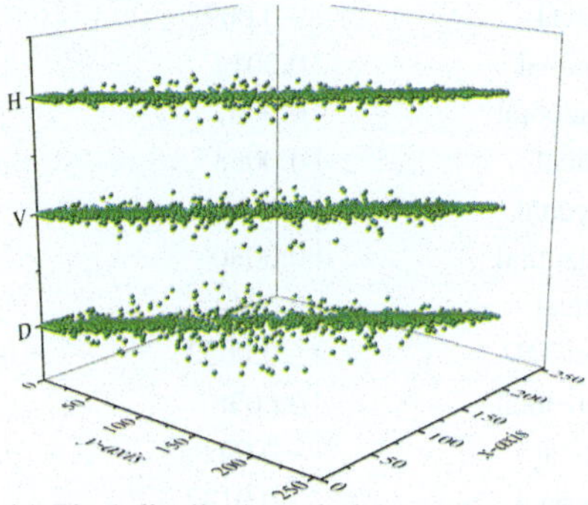

(c) Pixel distribution of plaintext image for green channel

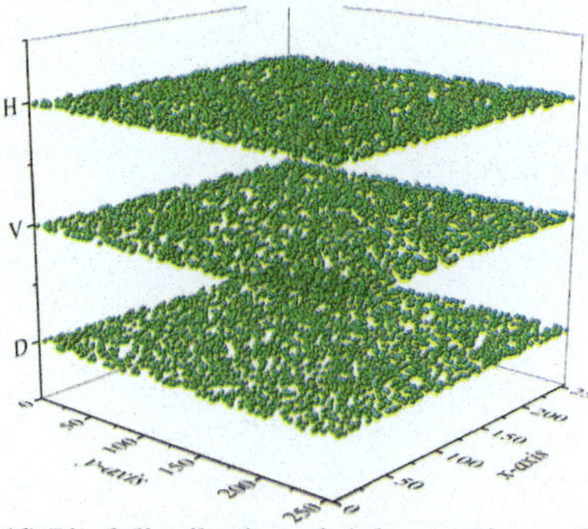

(d) Pixel distribution of cipher text image for green channel

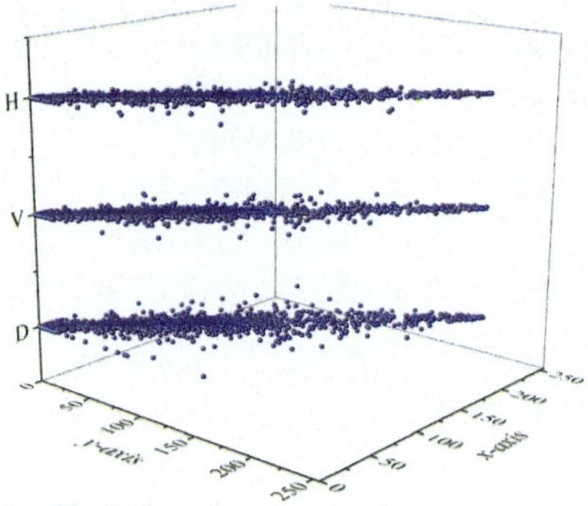

(e) Pixel distribution of plaintext image for blue channel

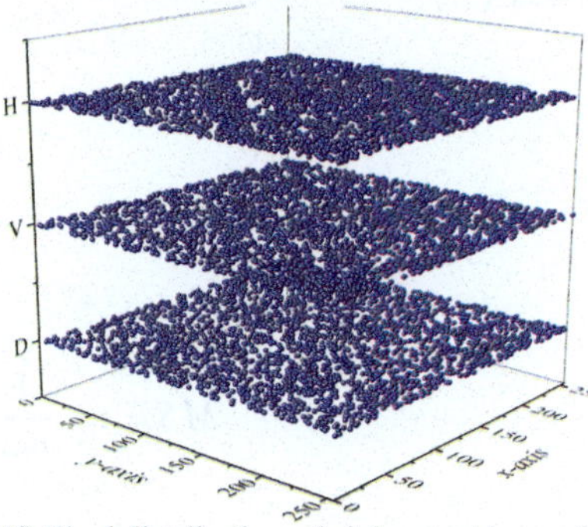

(f) Pixel distribution of cipher text image for blue channel

Fig. 4 Pixels Distribution for three directions: H, V, D for 4.2.07 image

Table 14 Comparison of correlation coefficients

	Channel	Direction	4.2.07 (512 × 512)
Our method	R	Horizontal	0.000475
	R	Vertical	−0.0000563
	R	Diagonal	0.000781
	G	Horizontal	−0.000793
	G	Vertical	0.000534
	G	Diagonal	0.000183
	B	Horizontal	−0.000589
	B	Vertical	0.000291
	B	Diagonal	−0.000593
Demirtas [15]	R	Horizontal	−0.0051
	R	Vertical	−0.0092
	R	Diagonal	0.0012
	G	Horizontal	0.0007
	G	Vertical	0.0068
	G	Diagonal	−0.0034
	B	Horizontal	0.0080
	B	Vertical	0.0014
	B	Diagonal	−0.0052
Li et al. [27]	R	Horizontal	0.0038
	R	Vertical	−0.0026
	R	Diagonal	−0.0187
	G	Horizontal	0.0014
	G	Vertical	0.0038
	G	Diagonal	0.0058
	B	Horizontal	−0.0094
	B	Vertical	0.0035
	B	Diagonal	0.0035

$$MSE := \frac{1}{mn}\sum_{i=1}^{m}\sum_{j=1}^{n}(e_{ij} - f_{ij})^2.$$

here, $E := (e_{ij})$ denotes the plaintext image, while $F := (f_{ij})$ represents the decrypted image obtained after the cropping process.

8.2.6 Noise Attacks

To compromise the integrity of ciphertext and impede successful decryption by the receiver, attackers often employ noise attacks. One prevalent form of noise attack

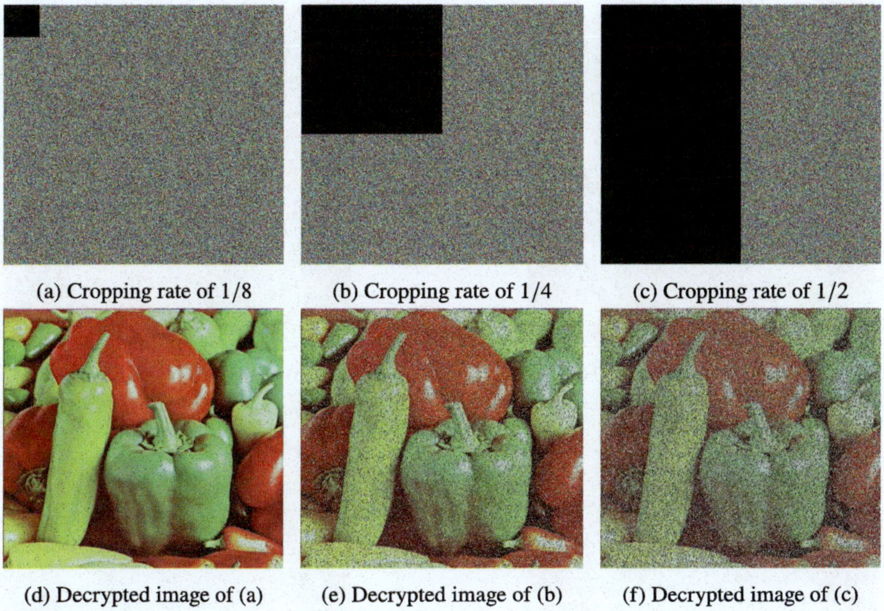

Fig. 5 Analysis of cropping attack for 4.2.07 image

Table 15 The PSNR values corresponding to the image that is decrypted under-cropping attacks

The Ratio of Crop	4.2.07
1/8	26.1022
1/4	14.1027
1/2	11.0987

is salt and pepper noise (SPN), where random black and white pixels are inserted into the ciphertext image. In our experiment, we analyze the impact of this attack by simulating various densities of SPN. Specifically, SPN densities of 0.01, 0.05, 0.1, and 0.2 are put into the encrypted image of 4.2.07. The results for the decryption can be observed in Fig. 6. Correspondingly, the PSNR for the deciphered images with SPN values of 0.01, 0.05, 0.1, and 0.2 are measured as 22.6490, 15.7299, 12.7526, and 9.8134, respectively. Figure 6 effectively showcases the resilience of the encryption and decryption method against this attack.

8.2.7 Analysis of Key Space

A brute-force attack is a widely used hacking technique where attackers attempt to crack passwords or encryption keys by systematically trying numerous possible combinations. In the context of image algorithms, a scheme with a short key lacks sufficient robustness to withstand exhaustive brute-force attacks. This means that if

(a) Decrypted image with 0.01

(b) Decrypted image with 0.05

(c) Decrypted image with 0.1

(d) Decrypted image with 0.2

Fig. 6 Images of cipher text with various densities of inserted salt and pepper noise (SPN)

the keyspace is relatively small, an attacker can iterate through all possible combinations until finding the correct one. However, the encryption method implemented with a key space that is larger than 2^{100} provides significantly higher security, making it highly resistant to brute-force attacks. With such an expansive key space, the number of possible keys becomes astronomically large, rendering it practically infeasible for an attacker to guess the correct key through brute force alone [28]. Our method utilizes a robust SHA 512-bit-length key that incorporates four floating-point numbers with an impressive precision of 10^{15}. These floating-point numbers play a vital role in determining the initial and control values for 2D-LSCM. By employing such a key, an extensive key space of $10^{(15\times4)} = 10^{60} \approx 2^{199}$ is achieved, surpassing the critical value by a significant margin. This vast keyspace demonstrates remarkable resilience against brute-force attacks, effectively safeguarding sensitive information.

8.2.8 Analysis of Key Sensitivity

The key sensitivity for the encryption method involves examining the effect of modifying just one bit in the original encryption key and subsequently decrypting the ciphered image using the modified key. This analysis enables the assessment of the encryption algorithm's resilience against minor key alterations. As depicted in Fig. 7, it is evident that decrypting the ciphered image with a key containing a single altered bit does not result in an accurate recovery of the original plain image. This observation highlights the strongness of the encryption method to even slight modifications in the encryption key, emphasizing the importance of precise key management and preservation for reliable and secure image decryption. To evaluate the key sensitivity of our method, a set of secret keys consisting of 512 bits in hexadecimal format is generated as follows:

Key_0=594E0E654EDF3B01B68E8E72669B939152524FBE53DB740C7EDD136
DFA96CE4167C2CBF1863651A0CE0C41C4D3719AB67FB4CB3ED5D467F206
8692286BE26622

Key_1=**A**94E0E654EDF3B01B68E8E72669B939152524FBE53DB740C7EDD136
DFA96CE4167C2CBF1863651A0CE0C41C4D3719AB67FB4CB3ED5D467F206
8692286BE26622

Key_2=594E0E654EDF3B01B68E8E72669B939152524FBE53DB740C7EDD136
DFA96CE4167C2CBF1863651A0CE0C41C4D3719AB67FB4CB3ED5D467F206
8692286BE2662**1**

Key_3=594E0E654EDF3B01B68E8E72669B939152524FBE53DB740C7EDD136
DFA96CE**5**167C2CBF1863651A0CE0C41C4D3719AB67FB4CB3ED5D467F206
8692286BE26622

To assess the key sensitivity of the method, the following variations of the secret key are generated: Key_1, Key_2, and Key_3. Key_1 is derived by flipping the first bit of Key, Key_2 by flipping the last bit, and Key_3 by flipping the 63rd bit. As illustrated in Fig. 7, it is apparent that despite differing from the original Key by only one bit, the decrypted images obtained using Key_1, Key_2, and Key_3 exhibit significant dissimilarities. This notable variation in the decrypted images highlights the effectiveness of the key to even minor modifications in the secret key. These findings underscore the critical significance of precise key management and preservation to uphold the integrity and security of encrypted images.

(a) Decryption by Key_0

(b) Decryption by Key_1

(c) Decryption by Key_2

(d) Decryption by Key_3

Fig. 7 Decryption with different keys

9 Discussions

In this section, we compare the results for the designed S-Box with different S-Boxes generated by various methods based on several evaluation criteria. The design S-Box achieves an impressive average non-linearity value of 107.50, surpassing the other S-Box generation methods in this regard. The design S-Box shows a BIC score of 0.4995, which indicates a good balance property. The design S-Box achieves an SAC score of 0.5039, suggesting that it passes the strict avalanche criterion and is capable of providing significant diffusion in cryptographic operations. The application of designed S-Box to image encryption demonstrates the effective utilization of the encryption technique to hide the content of the original image. Moreover, the original image was subsequently restored with exceptional accuracy using the appropriate decryption procedures.

References

1. Abbott TMC et al (2019) Dark energy survey year 1 results: constraints on extended cosmological models from galaxy clustering and weak lensing. Phys Rev D 99(12):123505. https://doi.org/10.1103/PhysRevD.99.123505
2. Ahmad M, Bhatia D, Hassan Y (2015) A novel ant colony optimization based scheme for substitution box design. Procedia Comput Sci 57:572–580. https://doi.org/10.1016/j.procs.2015.07.394
3. Ahmed HA, Zolkipli MF, Ahmad M (2019) A novel efficient substitution-box design based on firefly algorithm and discrete chaotic map. Neural Comput Appl 31(11):7201–7210
4. Azkaynak F (2020) On the effect of chaotic system in performance characteristics of chaos based s-box designs. Phys A 550:1–13
5. Babichev SA, Ries J, Lvovsky AI (2002) Quantum scissors: teleportation of single-mode optical states by means of a nonlocal single photon. Preprint at https://arxiv.org/abs/quant-ph/0208066v1
6. Basha SM, Mathivanan P, Ganesh AB (2022) Bit level color image encryption using Logistic-Sine-Tent-Chebyshev (LSTC) map. Optik 259:168956. https://doi.org/10.1016/j.ijleo.2022.168956
7. Ben Farah MA, Farah A, Farah T (2020) An image encryption scheme based on a new hybrid chaotic map and optimized substitution box. Nonlinear Dyn 99:3041–3064
8. Beneke M, Buchalla G, Dunietz I (1997) Mixing induced CP asymmetries in inclusive B decays. Phys Lett B 393:132–142 arXiv:0707.3168
9. Broy M (1992) Software engineering-from auxiliary to key technologies. In: Broy M, Denert E (eds) Software pioneers. Springer, New York, pp 10–13
10. Campbell SL, Gear CW (1995) The index of general nonlinear DAES. Numer Math 72(2):173–196
11. Cassal-Quiroga BB, Campos-Canton E (2020) Generation of dynamical S-boxes for block ciphers via extended logistic map. Math Probl Eng 2020:1–12
12. Çavuşoglu Ü, Zengin A, Pehlivan I, Kaçar S (2017) A novel approach for strong S-Box generation algorithm design based on chaotic scaled Zhongtang system. Nonlinear Dyn 87(2):1081–1094
13. Chen G (2008) A novel heuristic method for obtaining S-boxes. Chaos Solitons Fractals 36(4):1028–1036. https://doi.org/10.1016/j.chaos.2006.08.003
14. Chung ST, Morris RL (1978) Isolation and characterization of plasmid deoxyribonucleic acid from Streptomyces fradiae. In: Paper presented at the 3rd international symposium on the genetics of industrial microorganisms, University of Wisconsin, Madison, 4–9 June 1978
15. Demirtas M (2022) A new RGB color image encryption scheme based on cross-channel pixel and bit scrambling using chaos. Optik 265:169430. https://doi.org/10.1016/j.ijleo.2022.169430
16. Dey N (2018) Advancements in applied metaheuristic computing. IGI Global. https://doi.org/10.4018/978-1-5225-4151-6
17. Dey N, Ashour A, Bhattacharyya S (2020) Applied nature-inspired computing: algorithms and case studies. Springer
18. Dorigo M (1992) Optimization, learning and natural algorithms. PhD thesis, Politecnico di Milano, Italy
19. Dougherty ST, Sahinkaya S, Ustun D (2023) A novel method for image encryption using time signature-dependent s-boxes based on latin squares and the playfair system of cryptography. Multimed Tools Appl. https://doi.org/10.1007/s11042-023-15240-0
20. Farah T, Rhouma R, Belghith S (2017) A novel method for designing S-box based on chaotic map and teaching-learning based optimization. Nonlinear Dyn 88(2):1059–1074
21. Geddes KO, Czapor SR, Labahn G (1992) Algorithms for Computer Algebra. Kluwer, Boston
22. Hamburger C (1995) Quasimonotonicity, regularity and duality for nonlinear systems of partial differential equations. Ann Mat Pura Appl 169(2):321–354

23. Hao Z, AghaKouchak A, Nakhjiri N, Farahmand A (2014) Global integrated drought monitoring and prediction system (GIDMaPS) data sets. figshare. https://doi.org/10.6084/m9.figshare.853801
24. Hoseini R, Behnia S, Sarmady S, Fathizadeh S (2022) Construction of dynamical S-boxes based on image encryption approach. Soft Comput 26:13985–13997
25. Huynh-Thu Q, Ghanbari M (2008) Scope of validity of PSNR in image/video quality assessment. Electron Lett 44(13):800–801. https://doi.org/10.1049/el:20080522
26. Lambić D (2017) A novel method of S-box design based on discrete chaotic map. Nonlinear Dyn 87(4):2407–2413
27. Li Z et al (2020) A novel chaos-based color image encryption scheme using bit-level permutation. Symmetry 12(9):1497. https://doi.org/10.3390/sym12091497
28. Manjula G,Mohan H (2016) Constructing key dependent dynamic *S*-box for AES block cipher system. In: 2nd international conference on applied and theoretical computing and communication technology, pp 613–617. https://doi.org/10.1109/ICATCCT.2016.7912073
29. Seymour RS (1981) Conductive polymers. Plenum, New York
30. Shannon CE (1949) Communication theory of security systems. Bell Syst Tech J (Alcatel-Lucent) 28(4):656–715
31. Slifka MK, Whitton JL (2000) Clinical implications of dysregulated cytokine production. J Mol Med 78:74–80. https://doi.org/10.1007/s001090000086
32. Smith SE (1976) Neuromuscular blocking drugs in man. In: Zaimis E (ed) Neuromuscular junction. Handbook of experimental pharmacology, vol 42. Springer, Heidelberg, pp 593–660
33. Socha K, Dorigo M (2008) Ant colony optimization for continuous domains. Eur J Oper Res 185:1155–1173
34. Sosa PM (2016) Calculating non-linearity of Boolean functions with Walsh Hadamard transform. UCSB, Santa Barbara, pp 1–4
35. Stahl B (2020, Jun) deepSIP: deep learning of Supernova Ia Parameters. https://ui.adsabs.harvard.edu/abs/2020ascl.soft06023S
36. Tanyildizi E, Ozkaynak F (2019) A new chaotic S-box generation method using parameter optimization of one dimensional chaotic maps. IEEE Access 7:117829–117838
37. The USC-SIPI image database. https://sipi.usc.edu/database/
38. Tian Y, Lu Z (2016) S-box: six-dimensional compound hyperchaotic map and artificial bee colony algorithm. J Syst Eng Electron 27(1):232–241
39. Tian Y, Lu Z (2017) Chaotic S-box: intertwining logistic map and bacterial foraging optimization. Math Probl Eng :1-12. https://doi.org/10.1155/2017/6969312
40. Ullah A, Jamal SS, Shah T (2018) A novel scheme for image encryption using substitution box and chaotic system. Nonlinear Dyn 91(1):359–370
41. Wang Z, Bovik AC (2009) Mean squared error: love it or leave it? a new look at signal fidelity measures. IEEE Signal Process Mag 26(1):98–117. https://doi.org/10.1109/MSP.2008.930649
42. Wang Y, Wong KW, Li C, Li Y (2012) A novel method to design S-box based on chaotic map and genetic algorithm. Phys Lett A 376(6–7):827–833
43. Webster AF,Tavares SE (1986) On the design of S-boxes. In: Williams HC (ed) Advances in cryptology-CRYPTO-85 proceedings. CRYPTO 1985. Lecture notes in computer science, vol 218. Springer, Berlin, Heidelberg
44. Wu Y, Noonan JP, Agaian S (2011) NPCR and UACI randomness tests for Image encryption. Cyber J Multidiscip J Sci Technol J Sel Areas Telecommun 1:31–38
45. Zamli KZ, Kader A, Din F, Alhadawi HS (2021) Selective chaotic maps Tiki-Taka algorithm for the S-box generation and optimization. Neural Comput Appl 33:16641–16658
46. Zhang X, Gong Z (2022) Color image encryption algorithm based on 3D Zigzag transformation and view planes. Multimed Tools Appl. https://doi.org/10.1007/s11042-022-13003-x
47. Zhang Z, Tang J, Ni Hui, Huang T (2023) Image adaptive encryption algorithm using a novel 2D chaotic system. Nonlinear Dyn 11:10629–10652

Chapter 7
PID Controller Trained Using Ant Colony Algorithm for Load Frequency Control Problems on Three Equal-Area Interconnected Thermal Power with Renewable Energy Sources

D. Murugesan, K. Jagatheesan, Pritesh Shah, and Ravi Sekhar

1 Introduction

Power quality refers to the condition of electrical power given to the consumer, and it is essential for the proper and efficient operation of electrical and electronic devices. In order to provide the balance between the load and the generation in real time and guarantee a steady and dependable power supply, power systems use load frequency control (LFC). While LFC's primary goal is to maintain appropriate levels of frequency and tie-line power, it indirectly contributes to power quality improvement in the following ways: frequency regulation, voltage stability, reduction of power surges, mitigation of transients, and minimization of load shedding. It is important to note that while load frequency control can improve power quality indirectly, there are other specific power quality improvement techniques such as voltage regulation, harmonic mitigation, and power factor correction that address power quality

D. Murugesan (✉) · K. Jagatheesan
Paavai Engineering College, Namakkal 637018, TN, India
e-mail: benignsofts@gmail.com

K. Jagatheesan
e-mail: jaga.ksr@gmail.com

P. Shah · R. Sekhar
Symbiosis Institute of Technology, Symbiosis International (Deemed University), Pune 412115, India
e-mail: pritesh.shah@sitpune.edu.in

R. Sekhar
e-mail: ravi.sekhar@sitpune.edu.in

N. Dey (ed.), *Applications of Ant Colony Optimization and its Variants*, Springer Tracts in Nature-Inspired Computing, https://doi.org/10.1007/978-981-99-7227-2_7

issues more directly. Solar thermal, wind thermal, and fuel cell thermal power generation environments are the key proponents of carbon-free energy sources. Currently, the scarcity of power is increasing due to the sudden increase in the need for power from artificial intelligence-based modernized industry, population growth, entertainment, etc. Additionally, the high level of integration into the electrical grid presents difficulties, including variations in solar temperature and irradiance in addition variations in wind speed. These variations in voltage and frequency affect the system, which also widens the ongoing electricity supply and demand gap [1].

The control approach for energy management includes not only producing electricity from a power system based on renewable resources, but also sustaining the system voltage and frequency within the permitted ranges. Numerous study fields underline the need for additional clarification on the formulation, analysis, and evaluation of frequency inconsistency problems in the context of uncertainty. The producing poles in energy management always entail nonlinearities, temporal delays, and high order, which is the biggest issue with LFC-based proportional-integral-derivative (LFC-PID) controller parameter adjustment. In such a competitive environment, the LFC-PID should be successful in fulfilling the LFC's objectives. The PID regulator is useful in the LFC sector because of its benefits. However, optimizing the PID parameters in power systems based on renewable energy sources is a particularly challenging problem. According to the literature review, numerous researchers have used various controllers and optimization strategies to address the LFC. Researchers have proposed a wide variety of control strategies to preserve system stability by balancing power generation with load demand.

Nomenclature
$R_1 = R_2 = R_3 = R_4$ = Primary loop speed regulation = 2.4 Hz/puMW)
$B_1 = B_2 = B_3$ = Frequency Bias coefficient = 0.425 MW/Hz
$T_{12} = T_{23} = T_{31}$ = Synchronization time constant = 0.086 s
$T_{g1} = T_{g2} = T_{g3}$ = Mechanical governor time constant = 0.08 s
$T_{t1} = T_{t2} = T_{t3}$ = Steam turbine constant = 0.3 s
$T_{p1} = T_{p2} = T_{p3}$ = Generator time constants , respectivelys = 20 s
$K_{g1} = K_{g2} = K_{g3}$ = Governor gain value = 1
$K_{t1} = K_{t2} = K_{t3}$ = Steam Turbine gain value = 1
$K_{p1} = K_{p2} = K_{p3}$ = Generator gain value = 120
$T_{r1} = T_{r2} = T_{r3}$ = Reheat time constant = 10 s
$K_{r1} = K_{r2} = K_{r3}$ = Gain values of reheater = 0.5
K_{BESS} = Gain values of battery energy storage system = 0.0033
T_{BESS} = Time constant of battery energy storage system = 0.1 s
K_{FC} = Gain value fuel cell = 0.01
T_{FC} = Time constant of fuel cell = 0.4 s
K_{Solar} = Solar system gain value = 1
T_{Solar} = Solar system time constant = 1.5
T_{w1} = Wind turbine governor time constant = 6
T_{w2} = Wind turbine time constant = 0.041
K_{w1} = Wind turbine governor gain value = 1.25
K_{w2} =Wind turbine time constant = 1.4
J = Cost function
ΔP_D = Change in load in p.u.
$\Delta f_1 = \Delta f_2 = \Delta f_3$ = Deviation in frequency (Hz)
$D = \Delta P_D/\Delta f$ (p.u/Hz)
H = Inertia in each area
i = subscript related to the area
dij = i and j's distance from one another
P_{ij} = Towns i and j's chances of each other
$\alpha = \beta$ = determines the proportional interval between pheromone and heuristic values
Lk = length of Kth ant's journey
e(t) =Area control error signal
Q = pheromone quality-related constant
τij = Pheromone connected to the border between cities i and j
T_{sim} = Simulation time (sec)

Jagatheesan et al. [2] proposed a classical tuning method for a three-area conventional power framework (thermal + thermal + hydro) considering the generation rate constraint. Khuntia et al. [3] examined an ANFIS-based conventional three-area power system (thermal + thermal + hydro) with regard to the generation rate constraint. Golpira et al. [4] examined genetic algorithm-based integral controller design in a three-area thermal power framework in the existence of constraints. The triple-area thermal power system (hydro + nuclear + thermal) was evaluated by Chaturvedi et al. [5] using a GA algorithm-based fuzzy logic controller. An ant colony strategy is used in [6] to address the performance of the PIDD controller over the PID regulator in order to improve power quality. In [7], the TID controller examination of the LFC for a three unequal area network is analyzed in the existence of the DE. In [8, 9], a fuzzy PID regulator is developed for a three-area power system's LFC. For frequency deviation analysis, Murugesan et al. [10] addressed the FOPID regulator for single-area and double-area mutually dependent thermal power schemes. The authors of [11] used the dragonfly algorithm to create cascaded FOPI and FOPD regulators for LFCs in two-area and three-area power schemes. Meysam Gheisarnejad et al. [12] investigated the impact of a fuzzy fractional PID controller using the Cuckoo optimization method on frequency management for two- and three-area mutually dependent thermal power frameworks. According to Sasmita Padhy et al. [13], the performance of the multi-area power framework for autonomous generation control is being studied using the gray wolf optimization method for cascade

PI-PD regulator tuning. In [14], the authors looked at a three-area power structure configuration's resilient FOPID regulator performance mechanism. Three thermal networks with unequal areas were given load frequency control in [15], with the firefly method thought to perform better in the existence of nonlinearity. In [16], a cohort optimization-based PID regulator was used to assess a single-area nuclear power station for LFC. For the frequency control process of a double-area thermal power network with nonlinearity, the Cascaded structure of (1 + PID) PID was developed in [17].

From the literature, various nature-inspired methods have been developed to tune various regulator gain parameters in LFC/AGC problems for electrical power quality improvement. Examples include, a nature-inspired optimization method called teaching learning-based optimization (TLBO) [18, 19] simulates the process of teaching and learning among a group of people. It was proposed by Rao et al. in 2011 and is modeled after how students are taught and learn in a classroom. TLBO is a population-based method that locates the best answers to optimization issues in a variety of fields. Case study of nature-inspired optimization [20], Ant colony optimization (ACO) [21–23] is a well-liked metaheuristic method for optimization that was motivated by the foraging habits of ants looking for food. Marco Dorigo, who noted that genuine ant colonies display effective and adaptive behavior when seeking food sources and creating the best routes between their nest and food, first proposed it in the early 1990s. Another well-liked metaheuristic optimization technique, particle swarm optimization (PSO) [24], was influenced by the social behavior of fish schools and bird flocks. It was first presented as a population-based stochastic optimization technique by Eberhart and Kennedy in 1995. A metaheuristic optimization method inspired by nature, the bacterial foraging optimization algorithm (BFOA) [25], is based on how bacteria forage for resources in their surroundings. It was first up by Passino as a method of resolving optimization issues in 2002. A metaheuristic optimization method inspired by nature and the honey bees' feeding habits is called the artificial bee colony (ABC) [26]. Mimicking honey bees' foraging activity for food was first proposed by Karaboga in 2005 and is meant to solve optimization difficulties. Developed by R. Storn and K. Price in 1997, differential evolution [27] is a potent optimization algorithm. It is a member of the family of evolutionary algorithms that excels at addressing problems involving continuous optimization. A well-known metaheuristic optimization algorithm that was influenced by natural selection and evolution is the genetic algorithm (GA) [28]. It was developed by John Holland in the 1970s and is frequently used to address challenging optimization issues across numerous industries.

1.1 Analysis of the Gaps in the Literature

In this chapter, a thermal power system that includes thermal solar, thermal wind, and thermal fuel cells is put through the LFC using an ACO-tuned PID controller. The screening approach helped to identify the true research deficit in the field of LFC with renewable energy sources and extra power plants. In earlier research, non-renewable and renewable power plants were used both independently and as a component of an interconnected power network. However, in more recent times, a combination of renewable and non-renewable energy sources has been put into practice, allowing for continuous electricity production and use for power quality improvement, which offer more sustainable and environmentally friendly alternatives. The LFC research for a potential power plant is desperately needed. This chapter identifies and examines this gap.

1.2 Uniqueness of the Chapter

- Three area-connected conventional thermal power schemes are used to assess the capabilities of the artificial intelligence-based ACO approach by changing the pheromone, evaporation rate, iteration, and number of ants.
- In order to implement the LFC arrangement in the suggested system, a PID regulator is created.
- The ant colony process was successfully used to optimize the controller's gain values.
- Various arrangements (with and without renewable energy sources) are used to examine the achievement of the ACO.
- Additionally, the proposed scheme's ACO algorithm's transient performance is contrasted with a recently created soft computing technique.

1.3 Organization of the Chapter

The literature evaluation and introduction to the research work are covered in Sect. 1. The suggested system model and numerous power system equipment that have been investigated are provided in Sect. 2. The development of a suitable controller and objective function for an optimization approach is covered in Sect. 3. The use of ant colony optimization algorithms is given in Sect. 4. The suggested optimization approach is then used in Sect. 5 to introduce the expected system's conclusion. The chapter is created in accordance with the above structure.

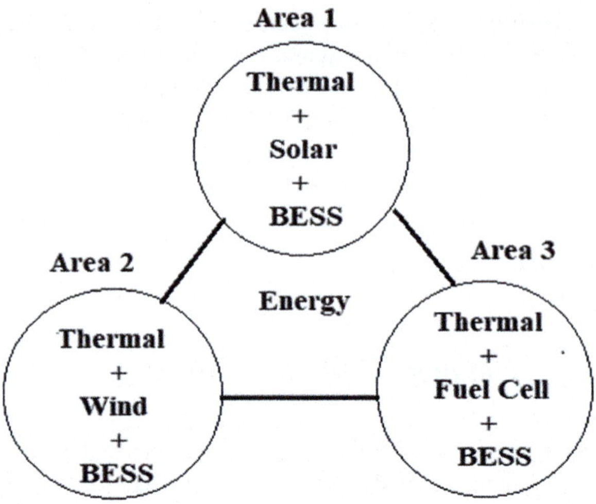

Fig. 1 Diagrammatic representation of the proposed work

2 System Under Study

2.1 Conventional Power System

MATLAB Simulink for LFC is used to model a triple-area thermal power scheme with equal-sized regions of 1: 2000, 2: 2000, and 3: 2000 MW for this article. Each of the three portions has a single reheat turbine installed, and AC tie lines link them together. MATLAB (R2014a) has been used to obtain dynamic responses for frequency changes, tie-line power flow out deviations, and area control error in all regions. The links between multiple power schemes make up the multi-area power network. Figure 1 depicts every component of the power system, including the speed-regulating system, turbine, reheater, and generator. Two outputs and three inputs are present in each section. The inputs consist of the tie-line power error (PTie), load disturbance (PD), and controller input (Pref; also known as u(t)). Area control error (ACE) and generator frequency (F) are the outputs. As described in Eqs. (1–3), the controllers receive the following sources of area control errors (ACE).

$$e_1(t) = ACE_1 = B_1\Delta F_1 + \Delta P_{tie\ i-j} \tag{1}$$

$$e_2(t) = ACE_2 = B_1\Delta F_1 + \Delta P_{tie\ i-j} \tag{2}$$

$$e_3(t) = ACE_3 = B_1\Delta F_1 + \Delta P_{tie\ i-j} \tag{3}$$

Transfer functions are applied to every component of the region to model it and to facilitate frequency-domain investigations. In Eqs. 4 and 5, the transfer function, represents the governor and the turbine, respectively:

$$G_T(s) = \frac{1}{1 + S\,T_t} \tag{4}$$

$$G_G(s) = \frac{1}{1 + S\,T_g} \tag{5}$$

The system's two inputs are ΔPref and ΔF, and its only output, ΔPG(s), is provided by Eq. 6.

$$\Delta P_G(s) = \Delta Pref - \frac{1}{R}\Delta F(s) \tag{6}$$

The transfer function in Eq. 7 represents the generator and load

$$G_P(s) = \frac{K_{PS}}{1 + S\,T_{PS}} \tag{7}$$

The generator load scheme has one output, ΔF(s) in Eq. 8, and two inputs, $\Delta P_T(s)$ and $\Delta P_D(s)$.

$$\Delta F(s) = G_P(s)\,[\Delta P_T(s) - \Delta P_D(s)] \tag{8}$$

2.2 *Renewable Power System*

A renewable power system, also known as a renewable energy system, refers to a collection of technologies and infrastructure that harness energy from renewable sources to generate electricity or power. Unlike fossil fuels (coal, oil, and natural gas), which are finite and contribute to environmental issues such as air pollution and climate change, renewable energy sources are sustainable, abundant, and have minimal impact on the environment. Here are some key components and sources commonly found in renewable power schemes.

A. Solar Power System

Solar energy systems, commonly referred to as solar power schemes, use solar energy to transform sunlight into electricity or heat. Due to their favorable effects on the environment, ability to produce their own energy, and declining costs, solar power schemes are growing in popularity. Photovoltaic (PV) systems and solar thermal frameworks are the two main categories of solar power schemes. In solar thermal systems, heat from the sun is used to warm a fluid (often water) and turn it into steam that powers a turbine to produce energy. The transfer function mode [6] for the solar system is mentioned in Eq. 9.

$$\text{Gsolar}(s) = \frac{K_{Solar}}{1 + ST_{solar}} \tag{9}$$

B. Wind Power System

A wind energy system uses the wind's kinetic energy to produce electricity. One of the most popular renewable energy sources, wind power has seen significant growth due to its advantages for the environment and declining costs. There are many different sizes of wind power systems, from modest household installations to massive utility-scale wind farms. From [19], the transfer function model for the wind system is taken into consideration.

C. Fuel Cell System

A fuel cell power system transfers chemical energy from a fuel into electrical energy, typically hydrogen, into electrical energy. It is a clean and efficient technology that offers an alternative to conventional combustion-based power generation methods. The idea behind fuel cells is that hydrogen and oxygen undergo a chemical reaction that results in the production of electricity, heat, and water as byproducts. Equation 10 indicates the fuel cell system transfer function model [20].

$$G_{Fuel}(s) = \frac{K_{FC}}{1 + ST_{FC}} \tag{10}$$

D. Battery Energy Storage System

The process of storing electrical energy in batteries for later use is known as a battery energy storage system (BESS). It is essential to contemporary power systems, helping to balance electricity supply and demand, manage grid stability, and integrate renewable energy sources more effectively. Battery energy storage systems can be deployed at various scales, ranging from small residential setups to huge utility-scale installations. The transfer function [20] model for the BESS is given in Eq. 11.

$$G_{BESS}(s) = \frac{K_{BESS}}{1 + ST_{BESS}} \tag{11}$$

3 Design of PID Controller/Objective Function

A PID regulator is commonly involved in load frequency control to adjust the output of power-generating units and maintain the system frequency. The controller receives input from frequency sensors that monitor the grid's actual frequency. It then compares this actual frequency with the nominal frequency (setpoint) and calculates the error (difference). The PID regulator applies three terms (proportional, integral, and derivative) to compute the appropriate control process and adjust the power production of the generators to sustain the system frequency at the desired value. Here is how each term is used in the Load Frequency Control scenario:

- Proportional (P) term: The P term of the PID controller responds to the instantaneous frequency error. It produces an output proportional to the difference between the actual frequency and the setpoint. This term is responsible for a rapid response to frequency deviations.
- Integral (I) term: The I term of the PID controller accumulates the frequency error over time. It helps eliminate any steady-state frequency offset that might arise due to load changes or other disturbances. The I term ensures that the average frequency error is reduced to zero over time.
- Derivative (D) term: The D term of the PID controller is fixed on the rate of change of the frequency error. It anticipates the future behavior of the system and helps in damping oscillations and reducing overshoot around the desired frequency.

The response of the PID regulator, the control action, is applied to the automatic generation control (AGC) system, which adjusts the power output of the generators accordingly. The goal is to sustain the electrical system frequency as near as possible to the setpoint, ensuring a stable and reliable power supply in the power grid. PID regulator are generally used: due to their simplicity, effectiveness, and ease of implementation in various power systems. However, in some advanced power systems, more complex and sophisticated control strategies may be employed, such as model-based predictive control or adaptive control, to address specific challenges and achieve better performance. The controller generates the control signal u(t) in Eq. 12, it can be stated as follows:

$$\mathrm{u(t)} = \mathrm{K_P e(t)} + \mathrm{K_I} \int e(t)dt + \mathrm{K_d}\frac{d}{dt}\mathrm{e(t)} \tag{12}$$

Integral time absolute error (ITAE) is a performance criterion used to assess the performance of a control system, especially in the context of tuning PID controllers or other feedback control systems. It is a measure of the cumulative absolute error over time, which quantifies how well the controlled variable follows the desired set point. The ITAE is described as follows:

$$\mathrm{J} = \int_0^\infty t\left|\left\{\Delta f_i + \Delta P_{\mathrm{tie}\,i-j}\right\}\right|dt \tag{13}$$

In Eq. 13, a cost function (J) is used to determine the controller gain values for Kp, Ki, and Kd. The ITAE considers both the magnitude of the error and the time it persists. It places more emphasis on larger errors and errors that persist for a longer duration. In control system design and PID controller tuning, the goal is to minimize the ITAE value, which indicates better system performance and faster response to setpoint changes. When tuning PID controllers using the ITAE criterion, the controller gains (Kp, Ki, and Kd) are adjusted systematically until the ITAE value is minimized. Several optimization techniques, such as trial-and-error, gradient-based methods, or advanced optimization algorithms, can be employed to find the optimal controller gains that lead to the lowest ITAE value. Keep in mind that while ITAE is a useful criterion for tuning controllers, and different control system applications.

4 Implementation of Ant Colony Optimization

The model of the arrangement under research is mentioned in Fig. 2, and the ACO code is written (in.mfile format). Initially, the conventional power system's population size (in terms of ants), cycle count, evaporation parameter, and pheromone strength were evaluated. Additionally, the PID regulator is taken into account for each area. The lower upper limits for KP, KI, and KD in this article are set at 0 and 1. A different software computes the goal function (ITAE) value for each population and transfers it to the SIMULINK.mdl file in the power system model through the workspace. Simulations were imposed in the MATLAB 32-bit (R2014a) environment on a Windows 10 computer with an AMD Pro A4-3350B running at 2.0 GHz and 4 GB of RAM. The algorithm is fixed on the following principles:

1. **Ant Agents**: In ACO, a Set of Artificial Ant Agents is Used to Explore the Solution Space. Each Ant Represents a Potential Solution to the Problem.

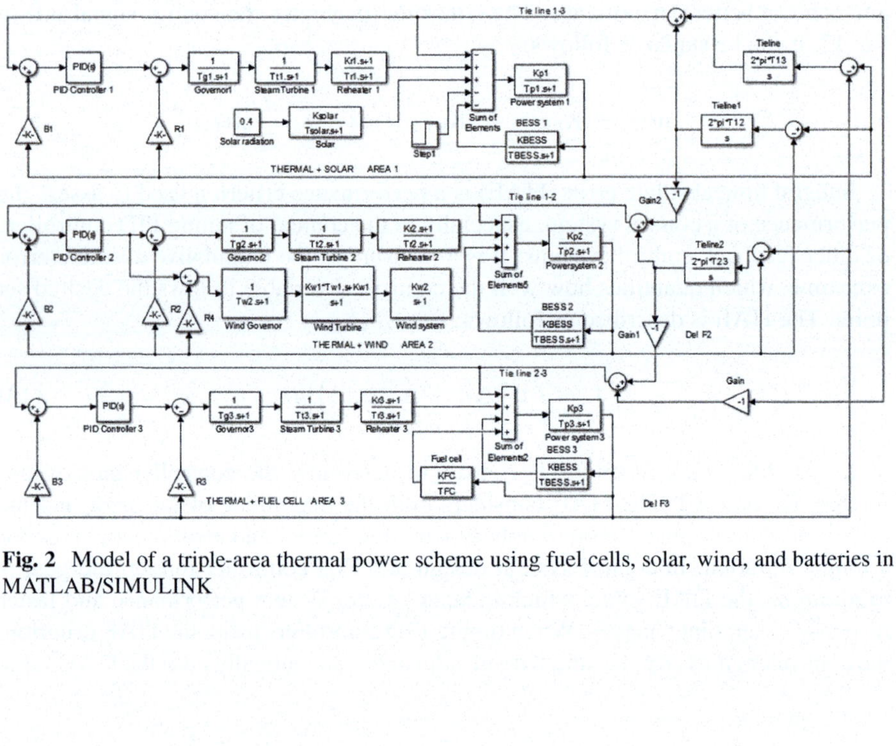

Fig. 2 Model of a triple-area thermal power scheme using fuel cells, solar, wind, and batteries in MATLAB/SIMULINK

2. **Pheromone Trails**: Ants produce a chemical material called "pheromone" on the roadway they traverse. Pheromones serve as a form of indirect communication between ants, allowing them to share information about the quality of their solutions.
3. **Exploitation–Exploration Balance**: Ants follow a balance between exploitation and exploration. Exploitation involves choosing paths with high pheromone levels (good solutions), while exploration allows ants to try new paths even with lower pheromone levels (potentially better solutions).

The main steps in the ant colony approach are as follows:

a. **Initialization**: Randomly place the ants on the problem's solution space, typically on different starting points.
b. **Solution Construction**: Each ant constructs a solution by iteratively choosing the next step based on a combination of pheromone levels and heuristic information. Heuristic information guides the ants toward potentially promising areas in the solution space.
c. **Solution Evaluation**: After constructing a whole solution, the ant evaluates its quality based on the cost function of the issues. For example, in the LFC, it would be the gain values of the controller, in the TSP, it would be the total distance traveled; in the VRP, it could be the total delivery cost.
d. **Pheromone Update**: Ants renew the pheromone matches on the paths they travel and fix the quality of their results. Typically, good solutions receive stronger pheromone deposits.
e. **Global Pheromone Update**: After all ants complete their planning phase, a global pheromone update is performed to evaporate existing pheromone levels and amplify the pheromone levels of good solutions.
f. **Termination**: For a certain number of cycles or until a termination condition is satisfied, the process of creating solutions, updating pheromone levels, and making global updates is duplicated.

Over time, the positive feedback from good solutions leads to the reinforcement of certain paths in the solution space. This, in turn, guides the ants to converge toward better solutions, effectively exploiting the knowledge gathered by the colony. These actual ant behaviors provided the motivation for numerous researchers [21–23] to develop the ACO method. ACO has been successfully applied to various optimizations, especially those with a combinatorial nature. It has shown promising results and often finds high-quality solutions in a reasonable amount of time. However, similar to other metaheuristic algorithms, ACO does not guarantee to find the global optimum solution and may become stuck in local optima for highly complex problems. Nevertheless, ACO remains a valuable tool for many real-world optimization challenges. Figure 3 shows the ACO algorithm's flowchart [23] to produce an optimal result. The starting parameters in the current study, including the quantity of ants, pheromone (τ), evaporation rate (ρ), and cycle, were chosen to produce improved results. The parameters chosen were 10/20/30/40/**50** ants, 0.2/0.4/**0.6**/0.8/1.0 pheromone, 0.55/0.65/0.75/0.85/**0.95** evaporation rate, and 20/40/60/**80**/100 iterations.

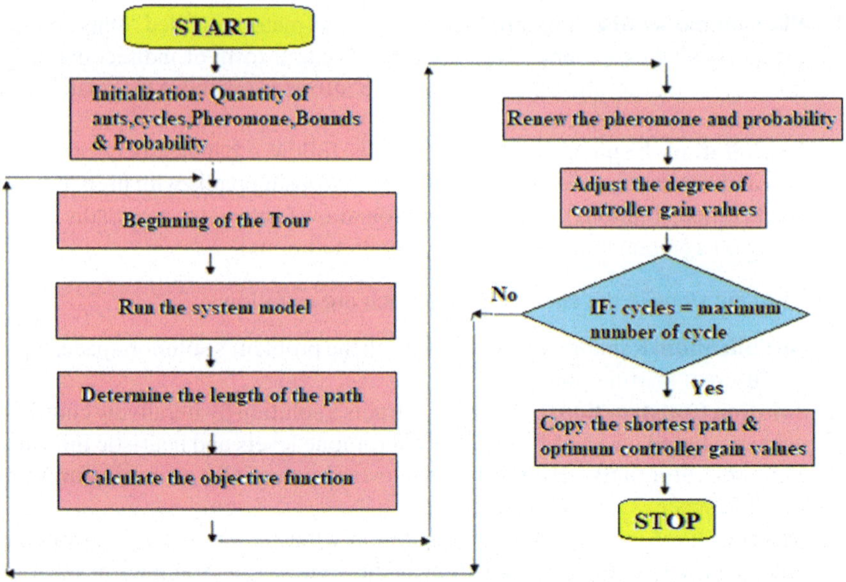

Fig. 3 Diagram illustrating the ACO algorithm

For the kth ant, the progress chances between towns i and j are as follows:

$$p_{ij}(t) = \frac{\tau_{ij}(t)^{\alpha}\left(\eta_{ij}\right)^{\beta}}{\sum_{j\varepsilon \text{nodes}} \tau_{ij}(t)^{\alpha}\left(\eta_{ij}\right)^{\beta}} \quad (14)$$

The following equation provides the parameter of pheromone information related to heuristic information:

$$\eta_{ij} = \frac{1}{d_{ij}} \quad (15)$$

The following global modification algorithm is provided once each ant begins its trip, and the pheromone is produced and changed on all fixed edges:

$$\tau_{ij}(t+1) = (1-\rho)\tau_{ij}(t) + \sum_{k \in \text{colony that used edge}(i,j)} \frac{Q}{L_K} \quad (16)$$

5 Results and Discussion

ACO applies to triple thermal power schemes that are linked together and both contain and exclude renewable energy sources. To investigate the importance of the ant colony algorithm, five simulation scenarios are executed. In scenario A, no renewable energy source is examined in the utilization of the LFC model. In scenario B, only solar power with a BESS with a base load is considered, and wind and fuel cells are ignored. In scenario C, a wind power scheme with a BESS with a base load is considered, while the solar plant and fuel cell are neglected. In scenario D, a fuel cell with a battery energy storage system with a base load is considered, and solar and wind plants are ignored. Exactly, in scenario E, all three renewable energy sources with battery energy storage systems with base loads are considered together and area control error is observed. For every scenario, a 0.01 p.u step load interruption is applied in area 1.

In scenario A, the ACO attempts to minimize the objective function while varying programmer design parameters such as pheromone, iterations, number of ants, and evaporation rate. Table 1 presents the most effective factors for pheromone variation. Table 1 clearly represents that area control errors are increased while changing pheromone values from 0.2 to 1.0. Additionally, observed for pheromone 0.6 based tuning gives improved result in minimizing settling time. Table 2 presents the most effective factors for iteration variation from 20 to 100. Based on Table 2, in 60 iterations, a minimum area control error (0.2864) is attained, and iteration 80-based tuning is seen to provide better results in lowering settling time. Additionally, Table 3 lists the most influential variables for ant variation from 10 to 50.

Considering Table 3, a low area control error (0.2891) is reached in 40 ants and the Lowering settling time is observed to be more successful with Ants 50-based adjustment. Table 4 includes a summary of the most significant variables affecting the evaporation rate in the range of 0.55 to 0.95. Table 4 shows that a 0.65 evaporation rate results in a low area control error (0.249), and that a 0.95 evaporation rate-based correction results in a lower settling time. This article investigates the performance of a triple-area power scheme utilizing user-defined ant colony algorithm parameters and renewable energy sources such as solar, wind, fuel cells, and battery energy

Table 1 Analysis of the settling time for a multi-area thermal power system when pheromone values are changed

User defined parameters		ΔF_1	ΔF_2	ΔF_3	$\Delta Ptie_{12}$	$\Delta Ptie_{23}$	$\Delta Ptie_{13}$	Area control error
Iteration (100), Ants (50), Evaporation parameter (0.95)	Pheromone							
	0.2	20	37	29	30	30	30	0.2817
	0.4	17	30	30	30	30	30	0.2873
	0.6	**17**	**27**	**28**	**30**	**30**	**30**	**0.2947**
	0.8	22	30	35	30	30	30	0.2884
	1	25	28	34	29	29	29	0.3343

Table 2 Analysis of the settling time for a multi-area thermal power system when iteration values are changed

User defined parameters		ΔF_1	ΔF_2	ΔF_3	$\Delta Ptie_{12}$	$\Delta Ptie_{23}$	$\Delta ptie_{13}$	Area control error
Pheromone (0.6), Ants (50), Evaporation parameter (0.95)	Iteration							
	20	22	31	31	31	31	30	0.3181
	40	17	37	31	29	29	29	0.3096
	60	20	29	29	30	30	30	0.2864
	80	**17**	**18**	**18**	**28**	**28**	**32**	**0.3117**
	100	17	32	32	30	30	30	0.3034

Table 3 Presents an analysis of the settling time for triple-area thermal power scheme

User defined parameters		ΔF_1	ΔF_2	ΔF_3	$\Delta Ptie_{12}$	$\Delta Ptie_{23}$	$\Delta Ptie_{13}$	Area control error
Pheromone (0.6), Iteration (80), Evaporation parameter (0.95)	Ants							
	10	19	27	28	30	30	30	0.2899
	20	22	35	29	30	30	30	0.2952
	30	24	37	32	31	31	31	0.3493
	40	22	35	35	30	30	30	0.2891
	50	**20**	**27**	**28**	**27**	**26**	**30**	**0.3**

storage. 50 ants are present, 80 iterations have been made, the evaporation parameter is fixed at 0.95, and the pheromone strength is 0.6. Table 5 shows the PID controller's gain values while employing ant colony optimization methods. Table 6 displays the gain parameters of PID regulator for (base load) traditional power systems using various optimization techniques.

By fine-tuning the controller gains, the power system can operate at a high level in schemes B, C, D, and E. Table 5 lists the calculated parameters that were most accurate. Figures 4, 5, 6, 7, 8 and 9 display the frequency inconsistency and net tie-line power transferred for every region for these values. It is clear that at steady-state, the LFC loop cancels the frequency deviations. Table 6 clearly demonstrates that the settling time of solar, wind, fuel cells, and solar + wind + fuel cells together is accomplished within 32 s, 38 s, 30 s, and 32 s, respectively, during the regulator design process. The optimized controller can handle the maximum allowable overshoot of 0.18 p.u, 0.0035 p.u, 0.003 p.u, and 0.18 by solar, wind, fuel cells, and solar + wind + fuel cells combined successfully as well.

Table 7 illustrates the controller gain values for three interconnected thermal power systems utilizing various optimization strategies, including GA, DE, PSO, TLBO, ACO, and CIO. For various optimization techniques offered, the most crucial time domain parameter in load frequency management, "Settling Time," is shown

Table 4 Analysis of the settling time for a triple-area thermal power system when evaporation parameters are changed

User defined parameters		ΔF_1	ΔF_2	ΔF_3	$\Delta Ptie_{12}$	$\Delta Ptie_{23}$	$\Delta Ptie_{13}$	Area control error
Pheromone (0.6), Iteration (80), Ants (50)	Evaporation parameters							
	0.55	22	28	28	30	30	30	0.2895
	0.65	22	28	28	30	30	30	0.249
	0.75	32	28	28	25	25	33	0.2818
	0.85	22	35	30	30	30	30	0.2879
	0.95	**17**	**27**	**27**	**30**	**30**	**30**	0.2947

Table 5 Gain values for the PID regulator while employing ant colony optimization methods

Various conditions	Proportional gain			Integral gain			Derivative gain		
	Kp1	Kp2	Kp3	Ki1	Ki2	Ki3	Kd1	Kd2	Kd3
Base load	0.93	0.56	0.56	0.98	0.57	0.57	0.31	0.6	0.6
Solar + BESS	0.98	0.99	0.2	0.35	0.99	0.27	0.35	0.99	0.27
Wind + BESS	0.95	0.65	0.97	0.73	0.77	0.66	0.78	0.75	0.32
Fuel Cell + BESS	0.93	0.98	0.31	0.98	0.77	0.6	0.98	0.77	0.6
Solar + Wind + Fuel Cell + BESS	0.82	0.28	0.78	1	0.99	0.91	0.2	0.52	0.26

Table 6 Time domain parameters of three areas interconnected power system with a renewable energy source

Method/ Triple areas	Peak overshoot (p.u Hz or MW)					Settling time in (s)				
	Scenarios					Scenarios				
	A	B	C	D	E	A	B	C	D	E
ΔF_1	0.001	0.18	0.0006	0.001	0.18	17	29	22	20	29
ΔF_2	0.001	0.18	0.0007	0.001	0.11	28	31	37	29	31
ΔF_3	0.001	0.18	0.0006	0.001	0.14	28	31	38	29	32
ΔP_{tie12}	0.003	0.01	0.0035	0.003	0.008	30	31	25	30	32
ΔP_{tie23}	0.003	0.01	0.0035	0.003	0.007	30	32	28	30	30
ΔP_{tie13}	0.0003	0.10	0.0004	0.0003	0.1	30	32	30	30	29

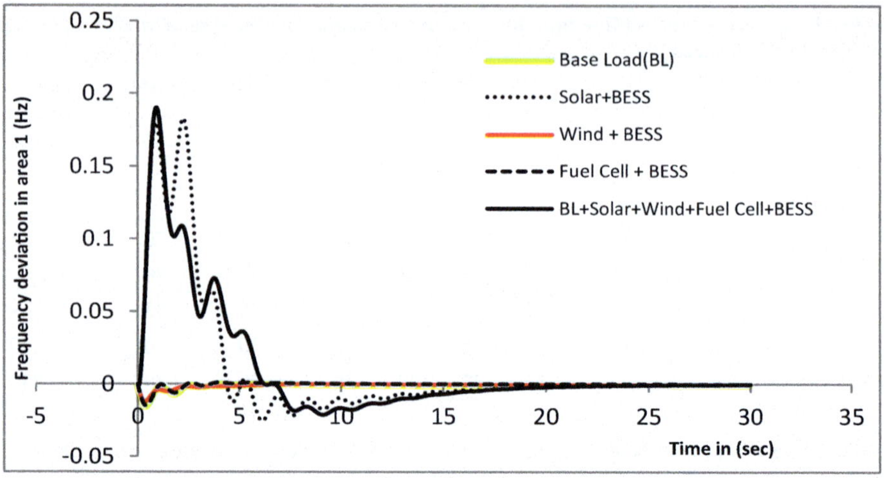

Fig. 4 Changes in frequency for three equal thermal power systems using renewable energy sources in area 2

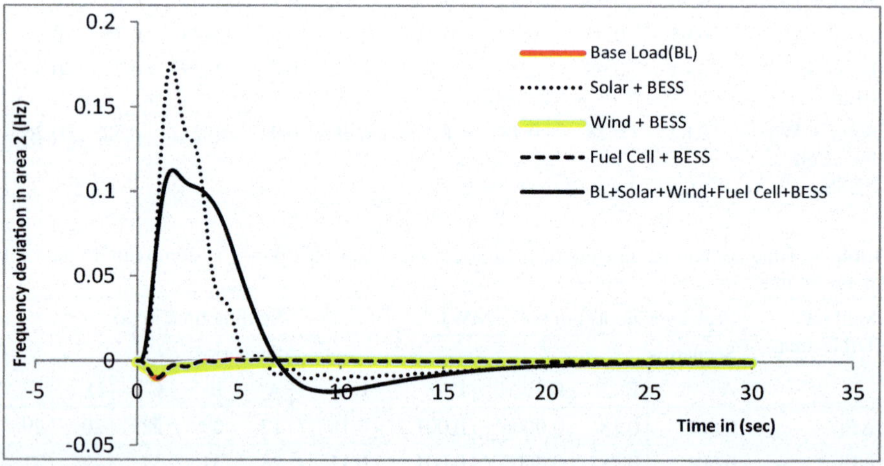

Fig. 5 Changes in frequency for three equal thermal power scheme using renewable energy sources in area 2

in Table 8. This makes it quite clear that the ACO-based PID controller response settles faster than the other methods. Tables 9 and 10 show the maximum overshoot and undershoot values for various optimization strategies. Figure 10 also provides a cylindrical chart analysis of the settling time of the three-area system for the various optimizations performed. Figures 11 and 12 also provide a cylindrical chart analysis of the overshoot and undershoot of a three-area system for various optimizations that are included.

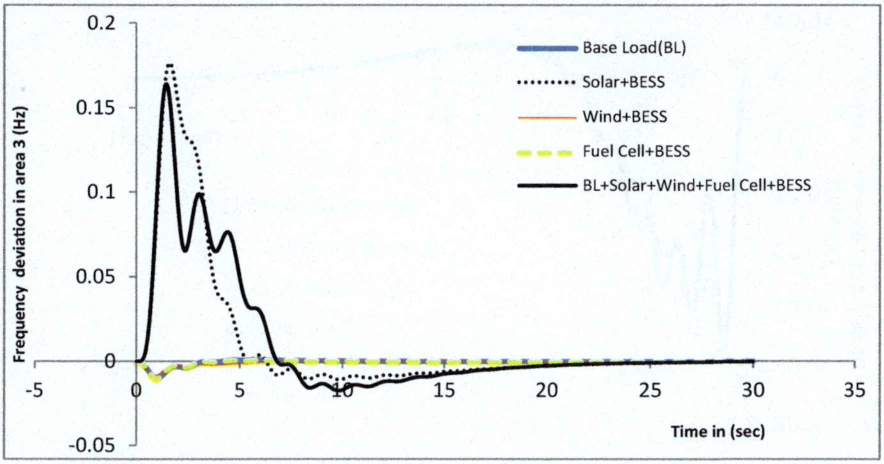

Fig. 6 Changes in frequency for three equal thermal power systems using renewable energy sources in area 3

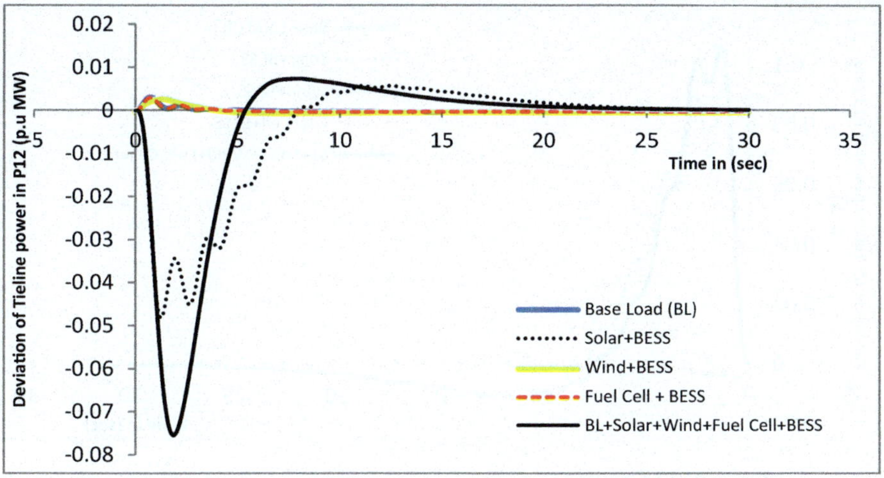

Fig. 7 Changes to the planned electrical power scheme's tie-line power 1–2

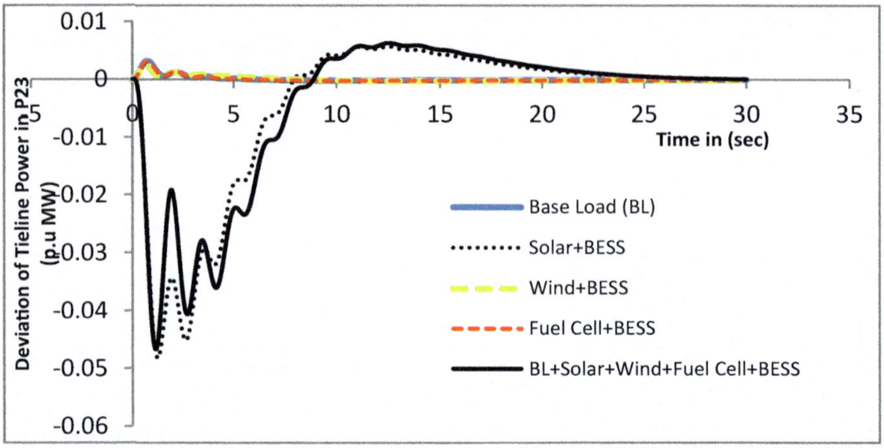

Fig. 8 Shows the variation in tie-line power 2–3 in three equal thermal power systems using renewable energy sources

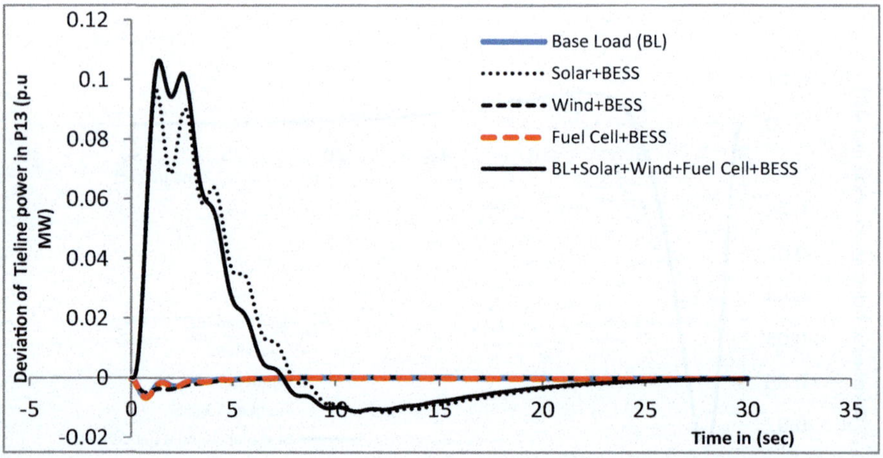

Fig. 9 Variation in tie-line power 1–3 in three equal thermal power systems with renewable energy sources

Table 7 Gain values of PID controllers for conventional power systems that use different optimization strategies

Gain	Proportional gain			Integral gain			Derivative gain		
Method	Kp1	Kp2	Kp3	Ki1	Ki2	Ki3	Kd1	Kd2	Kd3
ACO	0.93	0.56	0.56	0.98	0.57	0.57	0.31	0.6	0.6
CIO	0.96725	0.97986	0.2979	0.97375	0.91669	0.71255	0.22551	0.61376	0.76981
DE	0.97	1	0.92844	1	0.56901	0.88291	0.24088	1	0.85257
GA	0.99999	0.91404	0.62925	0.99966	0.99833	0.96644	0.31276	0.96349	0.53533
PSO	0.99918	0.99977	0.90304	0.99998	0.99972	0.98558	0.29686	0.60224	0.86054
TLBO	1	3.191e-14	0.00259	0.99491	0.9989	0.87085	0.4726	0.14704	1

Table 8 Shows the settling times for three equal-sized thermal power schemes that are mutually dependent on different optimization methods

Method/Triple areas	ACO	CIO	DE	GA	PSO	TLBO
ΔF_1	17	22	19	23	23	32
ΔF_2	27	30	27	28	35	35
ΔF_3	27	30	28	35	28	28
ΔP_{tie12}	30	32	32	30	30	30
ΔP_{tie23}	30	28	28	30	30	25
ΔP_{tie13}	30	30	30	30	30	32

Table 9 Shows the triple equal-area mutually dependent thermal power scheme's maximum overshoot for various optimization methods

Method/Triple areas	ACO $\times 10^{-3}$	CIO $\times 10^{-3}$	DE $\times 10^{-3}$	GA $\times 10^{-3}$	PSO $\times 10^{-3}$	TLBO $\times 10^{-3}$
ΔF_1	1.35	2	1	1.3	1.2	2
ΔF_2	0.9	1	0.75	0.75	0.75	1.2
ΔF_3	0.9	1	0.8	0.8	0.6	1.5
ΔP_{tie12}	3.2	3.4	3.25	3	3.25	3.2
ΔP_{tie23}	3.25	3.4	3.3	3.3	3.2	2.7
ΔP_{tie13}	0.25	0.25	0.25	0.25	0.25	0.5

Table 10 Shows the triple equal-area mutually dependent thermal power scheme's maximum undershoot for various optimization methods

Method/riple areas	ACO × 10^{-3}	CIO × 10^{-3}	DE × 10^{-3}	GA × 10^{-3}	PSO × 10^{-3}	TLBO × 10^{-3}
ΔF_1	14.7	15	15	14.2	14.5	13.5
ΔF_2	10	10.5	10	9	10	10
ΔF_3	10	10	10	10	9.5	8
ΔP_{tie12}	0.2	0.2	0.2	0.2	0.2	0.2
ΔP_{tie23}	0.2	0.2	0.2	0.2	0.2	0.25
ΔP_{tie13}	6.4	6.5	6.5	6.2	6.2	5.8

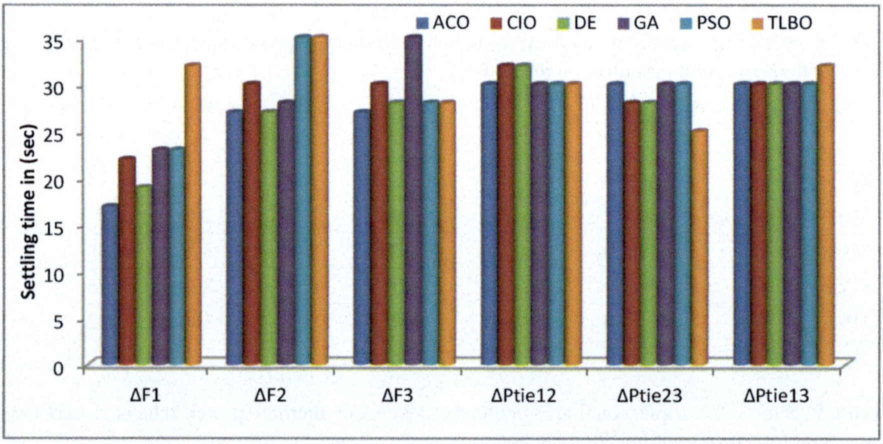

Fig. 10 Analysis of the settling time in three equal-area power systems using a cylindrical chart and several optimization methods

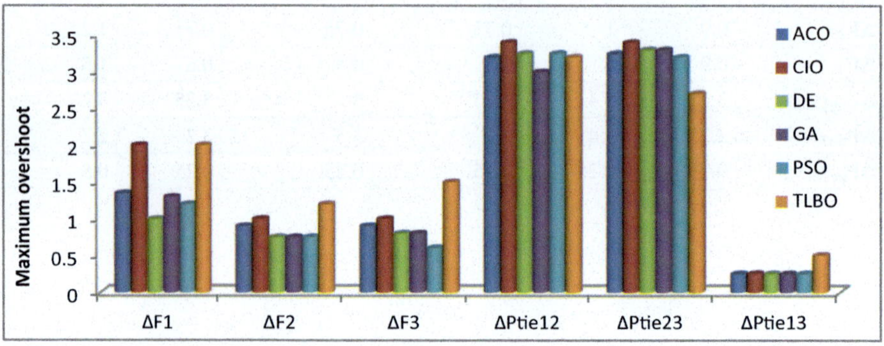

Fig. 11 Analysis of the overshoot in three equal area power systems using a cylindrical chart and several optimization methods

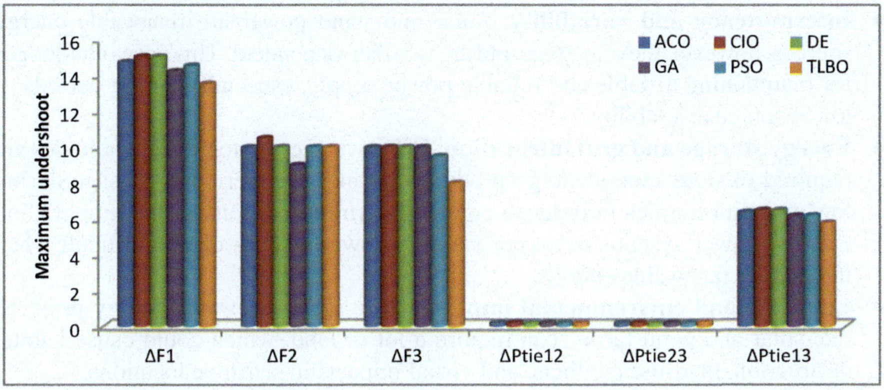

Fig. 12 Analysis of undershooting in three equal-area power systems using a cylindrical chart and several optimization methods

6 Conclusion

In this study, a proportional-integral-derivative regulator was used to examine the performance of a three-area linked thermal framework, and ITAE objective functions were used to identify the appropriate gain settings for the PID regulator. For the LFC, the effectiveness of three equal-area power plants with and without renewable energy sources was examined. The study also looked at changing an ant colony algorithm system design parameter. This information is used to generate the optimal optimization parameters in the ant colony method for the triple-area thermal power scheme with renewable energy sources. While integrating fuel cells, solar, wind, and BESS units with the suggested power system, no discernible changes in the settling time were found, but there were slight increases in oscillation, ranging from 0.001 to 0.18 p.u. Additionally, wind and fuel cell integration with the current power supply does not result in increased oscillation. Finally, it was determined that, in comparison to other computation techniques, the ant colony optimization-based controller equipped with an LFC system response for a triple-area power scheme has the least amount of waving and settling time.

7 Challenges and Restrictions

While renewable energy sources have numerous advantages, they also face several challenges and limitations that need to be addressed for their widespread adoption and integration into the global energy landscape. Some of the key challenges and limitations are:

- **Intermittency and variability**: Solar and wind power are renewable energy sources, however, they fluctuate and are weather-dependent. This poses challenges for maintaining a stable and reliable power supply, especially during periods of low resource availability.
- **Energy storage and grid integration**: Effective energy storage technologies are required to store excess energy produced during peak periods for usage during low generation times in order to combat intermittency. Furthermore, upgrading existing power systems to accept variable power sources can be difficult when integrating renewable energy.
- **Land use and environmental impact**: Large-scale renewable energy projects, like solar and wind farms, can require a lot of land, which could cause habitat destruction, land-use conflicts, and visual impacts in sensitive locations.
- **Initial high costs**: While the operating costs of renewable energy sources are generally low, the initial capital investment for setting up renewable energy infrastructure can be substantial, hindering their widespread adoption, especially in developing regions.
- **Resource availability and location**: The distribution of renewable energy resources is not uniform across the globe. For instance, areas with abundant sunlight or strong winds might not coincide with areas of high energy demand, leading to challenges in transporting energy from remote locations to urban centers.

Addressing these challenges requires a multi-faceted approach, including continued research and development to improve renewable energy technologies, advancements in energy storage solutions, supportive policies and incentives, grid modernization, and careful planning and community engagement for large-scale renewable projects. Overcoming these limitations will be crucial to accelerating the transition to a more sustainable and low-carbon energy future.

References

1. Kundur PS, Malik OP (2022) Power system stability and control. McGraw-Hill Education
2. Jagatheesan K, Anand B (2018) AGC of multi area hydro thermal power systems with GRC nonlinearity and classical controller. J Glob Inf Manag (JGIM) 26(3):11–24
3. Khuntia SR, Panda S (2012) Simulation study for automatic generation control of a multi-area power system by ANFIS approach. Appl Soft Comput 12(1):333–341
4. Golpira H, Bevrani H (2011) Application of GA optimization for automatic generation control design in an interconnected power system. Energy Convers Manage 52(5):2247–2255
5. Chaturvedi DK, Umrao R, Malik OP (2015) Adaptive polar fuzzy logic based load frequency controller. Int J Electr Power Energy Syst 66:154–159
6. Murugesan D, Jagatheesan K, Boopathi D (2021) Meta-heuristic strategy planned controller for frequency supervision of integrated thermal plant with renewable source. In: 2021 IEEE 3rd PhD colloquium on ethically driven innovation and technology for society (PhD EDITS). IEEE, pp 1–2

7. Sahu RK, Panda S, Biswal A, Sekhar GC (2016) Design and analysis of tilt integral derivative controller with filter for load frequency control of multi-area interconnected power systems. ISA Trans 61:251–264
8. Sahu RK, Panda S, Pradhan PC (2015) Design and analysis of hybrid firefly algorithm-pattern search based fuzzy PID controller for LFC of multi area power systems. Int J Electr Power Energy Syst 69:200–212
9. Sahu BK, Pati S, Panda S (2014) Hybrid differential evolution particle swarm optimisation optimized fuzzy proportional–integral derivative controller for automatic generation control of interconnected power system. IET Gener Transm Distrib 8(11):1789–1800
10. Murugesan D, Jagatheesan K, Shah P, Sekhar R (2023) Fractional order PIλDμ controller for microgrid power system using cohort intelligence optimization. Results Control Optim 11:10021
11. Çelik E (2021) Design of new fractional order PI–fractional order PD cascade controller through dragonfly search algorithm for advanced load frequency control of power systems. Soft Comput 25(2):1193–1217
12. Gheisarnejad M, Khooban MH (2019) Design an optimal fuzzy fractional proportional integral derivative controller with derivative filter for load frequency control in power systems. Trans Inst Meas Control 41(9):2563–2581
13. Padhy S, Panda S, Mahapatra S (2017) A modified GWO technique based cascade PI-PD controller for AGC of power systems in presence of plug in electric vehicles. Eng Sci Technol Int J 20(2):427–442
14. Delassi A, Arif S, Mokrani L (2018) Load frequency control problem in interconnected power systems using robust fractional PIλD controller. Ain Shams Eng J 9(1):77–88
15. Padhan S, Sahu RK, Panda S (2014) Application of firefly algorithm for load frequency control of multi-area interconnected power system. Electr power Comp Syst 42(13):1419–1430
16. Murugesan D, Jagatheesan K, Kulkarni AJ, Shah P (2023) A socio inspired technique in nuclear power plant for load frequency control by using cohort intelligence optimization-based PID controller. In: Renewable energy optimization, planning and control: proceedings of ICRTE 2022. Springer Nature Singapore, Singapore, pp 1–12
17. Murugesan D, Shah P, Jagatheesan K, Sekhar R, Kulkarni AJ (2022) Cohort intelligence optimization based controller design of isolated and interconnected thermal power system for automatic generation control. In: 2022 second international conference on computer science, engineering and applications (ICCSEA). IEEE, pp 1–6
18. Murugesan D, Jagatheesan K, Kulkarni AJ, Shah P, Sekhar R (2023) Comparison of biologically inspired algorithm with socio inspired technique on load frequency control of multi-source single-area power system. In: Applied genetic algorithm and its variants: case studies and new developments). Springer Nature Singapore, Singapore, pp 185–208
19. Rao RV, Savsani VJ, Vakharia DP (2012) Teaching–learning-based optimization: an optimization method for continuous non-linear large scale problems. Inf Sci 183(1):1–15
20. Dey N, Ashour AS, Bhattacharyya S (eds) (2020) Applied nature-inspired computing: algorithms and case studies. Springer, Singapore
21. Dorigo M, Di Caro G (1999) Ant colony optimization: a new meta-heuristic. In: Proceedings of the 1999 congress on evolutionary computation-CEC99 (Cat No 99TH8406), vol 2. IEEE, pp 1470–1477
22. Dorigo M, MB S (2006) T: ant colony optimization-artificial ants as a computational intelligence technique. Universit Libre de Bruxelles. Technical report TR/IRIDIA/2006–023
23. Jagatheesan K, Anand B, Dey KN, Ashour AS, Satapathy SC (2018) Performance evaluation of objective functions in automatic generation control of thermal power system using ant colony optimization technique-designed proportional–integral–derivative controller. Electr Eng 100:895–911
24. Gozde H, Taplamacioglu MC (2011) Automatic generation control application with craziness based particle swarm optimization in a thermal power system. Int J Electr Power Energy Syst 33(1):8–16

25. Ali ES, Abd-Elazim SM (2011) Bacteria foraging optimization algorithm based load frequency controller for interconnected power system. Int J Electr Power Energy Syst 33(3):633–638
26. Naidu K, Mokhlis H, Bakar AA (2014) Multiobjective optimization using weighted sum artificial bee colony algorithm for load frequency control. Int J Electr Power Energy Syst 55:657–667
27. Mohanty B, Panda S, Hota PK (2014) Controller parameters tuning of differential evolution algorithm and its application to load frequency control of multi source power system. Int J Electr Power Energy Syst 54:77–85
28. Murugesan D, Jagatheesan K, Shah P, Sekhar R (2023) Optimization of load frequency control gain parameters for stochastic microgrid power system. J Robo Control (JRC) 4(5):726–742

Chapter 8
Antlion-Facing Ant Colony Optimization in Parameter Identification of the MR Damper as a Semi-active Control Device

Salar Farahmand-Tabar and Sina Shirgir

1 Introduction

Optimization algorithms are crucial in diverse fields, as they aim to find the best problem-solving approach. However, each algorithm has its strengths and limitations. While certain algorithms excel in particular problem types, they may fall short in others, necessitating improvements and adjustments to enhance their dependability and effectiveness [1, 2]. For example, gradient descent algorithms can quickly converge to local minima but struggle to find global optima. Conversely, metaheuristic methods such as genetic algorithms and particle swarm optimization are skilled at locating global optima by exploring the search space, even though they may require modifications and take longer to converge [3–9].

Researchers frequently combine multiple techniques such as hybridization, parallelization, and parameter tuning to bolster the resilience and efficiency of optimization algorithms. Hybrid algorithms merge two or more optimization techniques, capitalizing on their strengths while compensating for weaknesses. Parallelization involves running the algorithm on multiple processors to expedite the optimization process. Parameter tuning entails adjusting the algorithm's parameters for better performance on specific problems. Ultimately, the pursuit of improving optimization algorithms stems from the desire to discover more efficient and effective solutions to problems [10–16].

S. Farahmand-Tabar (✉)
Department of Civil Engineering, Faculty of Engineering, University of Zanjan, Zanjan, Iran
e-mail: farahmandsalar@znu.ac.ir

S. Shirgir
Department of Structural Engineering, Faculty of Civil Engineering, University of Tabriz, Tabriz, Iran

N. Dey (ed.), *Applications of Ant Colony Optimization and its Variants*, Springer Tracts in Nature-Inspired Computing, https://doi.org/10.1007/978-981-99-7227-2_8

In the realm of optimization algorithms, nature-inspired methods have gained considerable attention due to their ability to tackle complex problems with efficiency and effectiveness. Two such algorithms, the antlion optimizer (ALO) [17] and ant colony optimization (ACO) [18], are inspired by the ants' foraging behavior and the social behavior of ant colonies, respectively. Combining these two algorithms into a hybrid approach holds the promise of leveraging their complementary strengths to enhance the optimization process and address the limitations present in each individual algorithm. This chapter explores the principles, motivations, and potential advantages of the hybridization of ALO and ACO and its implications in solving real-world optimization problems.

The Antlion Optimizer is a population-based optimization method inspired by the hunting strategy of antlions (an insect species known for constructing conical pits to trap ants). In ALO, the population of antlion agents performs a random walk and leaves pheromone trails behind. These pheromones guide the movement of the agents, attracting them toward regions with higher pheromone concentrations. As the algorithm progresses, the pheromone concentrations increase around promising solutions, encouraging exploration and exploitation of the search space. The primary advantage of ALO lies in its efficient exploration ability, enabling it to locate potential solutions in large and complex optimization landscapes. However, ALO might suffer from premature convergence, wherein the algorithm settles too quickly on local optima, limiting its global search capability.

Ant colony optimization is another nature-inspired optimization algorithm based on the foraging behavior of ants. ACO employs a population of artificial ants that traverse the problem domain, laying down pheromone trails that indicate the desirability of a particular path. The positive feedback loop of pheromone reinforcement allows ACO to focus on paths with higher pheromone levels, effectively guiding the search toward promising solutions. Over time, the algorithm tends to converge to the best solution discovered so far. ACO's major strength lies in its ability to exploit promising regions of the search space, making it particularly proficient in local search and refinement. However, ACO might struggle when dealing with large-scale problems or problems with complex, rugged landscapes, as it can become trapped in local optima and may require additional mechanisms to maintain diversity in exploration.

The hybrid approach is expected to excel in solving complex real-world optimization problems, such as engineering design, logistics, financial modeling, and many others. The combination of exploration and exploitation capabilities is anticipated to enable the hybrid algorithm to efficiently navigate through intricate search spaces, discovering high-quality solutions in a more reliable and timely manner.

This chapter introduces a robust hybrid algorithm named the antlion-facing ant colony optimizer (AFACO). AFACO combines the strong search capabilities of ants from ant colony optimization (ACO) with the effective hunting abilities of antlions. By adopting the discrete movement approach of ants, AFACO navigates paths based on pheromone values on edges while still retaining the antlion's predation process. To further enhance its performance when encountering local optima, AFACO incorporates a mutation operator. Through extensive experiments conducted on parameter identification of the MR damper as a semi-active control device in structures, AFACO demonstrates exceptional optimization capabilities, surpassing other considered methods.

2 Related Works on Parameter Identification

Hysteresis behavior, characterized by its complexity and nonlinearity, lacks a comprehensive mathematical framework to accurately represent it. Nevertheless, specific cases of hysteresis systems are described using descriptive models that adhere to the governing physical principles of the problem. One notable example is the Bouc-Wen model, first introduced by Bouc [19] and further developed by Wen [20–22], which aims to obtain the general characteristics of hysteresis. The Bouc-Wen model is known for its versatility in generating various hysteresis patterns, gaining popularity, and finding extensive application in diverse engineering problems [23–28].

For the precise representation of hysteresis behavior in magnetorheological (MR) dampers, several models have been developed [29–35]. Among these models, the modified Bouc-Wen model has gained significant attention and acceptance due to its effectiveness in capturing the responses observed in real MR dampers. Within the Bouc-Wen model, the system's deformation is described by a first-order nonlinear differential equation, with certain parameters associated with the nonlinear restoring force remaining undefined. By carefully selecting appropriate parameter values, the Bouc-Wen model's response closely aligns with the actual behavior of the nonlinear system.

The procedure of discovering the most suitable values for the parameters is commonly known as "*Parameter Identification*" or "*System Identification*", and it involves solving an optimization problem. Various methods can be employed for this purpose [36, 37]. For example, a hybrid particle swarm optimization approach was successfully applied [38] to obtain the optimal parameter set for the Bouc-Wen model, accurately reproducing the hysteresis behavior of a steel cantilever beam observed from the experiment. Kwok et al. [39] used the genetic algorithm to identify the Bouc-Wen relationships that describe the hysteretic nonlinear behavior of MR dampers. Similarly, Talatahari et al. [40] and Liu et al. [41] applied the adaptive charged system search and simulated annealing algorithm, respectively, to effectively determine the optimal Bouc-Wen model parameters for MR dampers. Additionally, Talatahari and

Mohajer [42] introduced the enhanced imperialist competitive algorithm as a solution to this optimization problem.

In some studies, based on reliability analysis, Shirgir et al. presented a simplified model for the Buoc-Wen model and identified its parameters using metaheuristic strategies [43, 44]. The behavior of MR dampers has also been considered from different perspectives. Raeesi et al. designed MR dampers using metaheuristic methods [45]. Additionally, in a study, the reliability of MR damper performance was evaluated under uncertainty [46].

3 Optimization Method

Metaheuristic optimization faces a significant challenge in maintaining a diverse population of candidate solutions during the search process. This approach is vital to prevent premature convergence and increase the likelihood of finding high-quality solutions. In recent decades, there has been a remarkable increase in interest in developing techniques that enhance the search strategy within metaheuristics, with a specific focus on improving the efficiency and effectiveness of these methods.

In the subsequent sections of this chapter, the intricacies of the hybrid algorithm are delved into, with the methods of combining ALO and ACO detailed, an updating mechanism for pheromones and search agents proposed, and comprehensive experiments conducted to evaluate its performance against real-world problems.

3.1 Antlion Search Strategy

The antlion optimizer (ALO) is a cutting-edge optimization method that draws inspiration from both artificial intelligence and swarm intelligence principles. It is inspired by the hunting behaviors of antlions during their larval stage. In the ALO, particles are utilized to represent antlions and ants, simulating the interactions observed in real-life antlions that set traps underground to catch ants (as depicted in Fig. 1). The core focus of ALO lies in the interactions between these particles, where ants move within the search space, and antlions strive to capture the ants through their traps [17, 47].

Fig. 1 The hunting behavior of antlions using cone-shaped traps for ants

As the optimization process unfolds, the antlions improve their trapping techniques, becoming more skillful at their task. To model the ants' movement, the ALO adopts the "*random walk*" approach, altering the ants' positions as parameters for the objective function(s) to discover optimal values [17, 47–49]. By combining these concepts, the ALO aims to achieve efficient and effective optimization, drawing inspiration from the natural world.

ALO has been extensively studied in various sources [47–51]. In this approach, ants explore the search space by performing random walks that encompass all dimensions of the problem. During ALO iterations, the ants' random walks are influenced by traps set by antlions. These traps, known as pits, are created by antlions according to their fitness levels, and larger pits have a higher probability of capturing ants. Surprisingly, even the most skilled ant can be ensnared by an antlion during the ALO process.

The movement of ants toward antlions is facilitated by dynamically reducing their range of random walks. In the event that an ant is discovered to be more fit than an antlion, it becomes captured and concealed beneath the sand by the antlion. During ALO iterations, the selection of antlions to be processed follows the roulette wheel method, commonly employed in genetic algorithms. To increase the probability of catching other prey, antlions are relocated to the position of the last captured ant, and a new pit is constructed at that location. Here is a pseudocode (Algorithm 1) representation of the standard ALO, along with the essential equations [17, 47–49]:

Algorithm 1. Pseudocode of ALO

Step 1 (Initialization Phase): Distribute a set of M ants and N ant-lions randomly across the search space.

Step 2: Evaluate the fitness of all antlions and ants by applying the relevant fitness function. Accordingly, the best antlion is identified.

Step 3: Continue performing subsequent steps until the stopping criteria are satisfied.

I: For every ant, carry out the following sub-steps:

a: Utilize the roulette wheel selection method to pick an ant-lion for each ant.

b: Apply the given equations to modify the min. and max. values (c and d) based on the ratio (I):

$$c^{time} = c^{time}/I\ ; d^{time} = d^{time}/I \tag{1}$$

c: Conduct a random walk and then normalize the results:

$$X(time) = [0, \ldots cs(2r(time_1) - 1), cs(2r(time_2) - 1), \ldots, cs(2r(t_n) - 1)] \tag{2}$$

where cs is the cumulative sum,

and $r(time) = \begin{cases} 1, \ random\ number > 0.50 \\ 0, \ random\ number \leq 0.50 \end{cases}$

$$X_i^{time} = \frac{(X_i^{time} - a_i) \times (di - c_i^{time})}{(d_i^{time} - a_i)} + c_i \tag{3}$$

d: Update the position of the relevant ant through the given equation:

$$AntPosition_i^{time} = \frac{R_{AntLion}^{time} - R_{BestAntLion}^{time}}{2} \tag{4}$$

where $R_{AntLion}^{time}$ and $R_{BestAntLion}^{time}$ are the random walk around the selected and best antlion with the roulette wheel, respectively.

II: Compute the fitness values for each of the ants.

III: Substitute an antlion with related ant if the ant is found to be more fit, applying the following equation:

$$AntLionPosition_j^{time} = AntPosition_i^{time} \tag{5}$$

IV: If an ant-lion is superior to the current best antlion, update the best antlion with the improved one.

Step 4: The best antlion represents the optimal solution.

The procedure can be technically described, as documented in reference [52]. The ALO begins with the random initialization of matrices representing both ants and antlions. In each iteration, the positions of ants corresponding to antlions are determined using the roulette wheel technique. Simultaneously, the best antlion encountered thus far is continually updated based on its relation to the current iteration number. This updating process modifies the position of the best antlion through two random walks, one centered around the selected best ant and the other around the antlion. During each random walk, the points are evaluated using the fitness function.

If an ant is discovered to be superior to other antlions, its position is substituted with that of the antlions in the following iteration. Additionally, the best antlion is compared with the optimum one obtained during the optimization process, and if required, it is replaced at its original location.

3.2 *Search Strategy of Ants in the Colony*

Ant colony optimization (ACO) is a bio-inspired metaheuristic that imitates ants' foraging behavior to discover food sources [18]. What sets ACO apart from other metaheuristics is its ability to support parallel processing, operate independently of the process, and provide valuable feedback on the behaviors of ants within the search space [53].

Ants, while searching for food, display remarkable navigation skills, efficiently exploring the shortest route between the food source and their nest. This efficiency is due to their practice of leaving a chemical substance known as pheromone along their paths. Pheromone acts as a means of communication among ants, indicating the shortest path to the food source. As ants travel, they sense the presence of pheromone left by other ants, guiding them to prefer paths with higher pheromone concentrations, thus increasing the likelihood of more ants following the same path.

ACO relies on two essential components—the heuristic information and pheromone trail—to make decisions in probabilistic situations. As ants move along a path, they update the pheromone level on each feature. Paths with elevated pheromone concentrations signify a greater probability of belonging to the shortest path, as more ants are inclined to follow those routes. Consequently, the path with the highest pheromone level becomes the most alluring option and is selected by a larger number of ants as the shortest route.

The process begins by assigning a pheromone value $\tau_0 = 1$ to all M features, and ants are distributed randomly across a set of solutions for a predetermined number of generations T. At each iteration g, the transition probability $TP_i^k(g)$ for the k th ant at the i th variable is calculated following the guidelines provided in references [54, 55].

$$TP_i^k(g) = \begin{cases} \frac{[\tau_i(g)]^\alpha . [\eta_i]^\beta}{\sum f_{j \notin j_i^k} [\tau_j(g)]^\alpha . [\eta_j]^\beta} & f_{j \notin j_i^k} \\ 0 & otherwise \end{cases} \tag{6}$$

The ants' exploration in ACO depends on the set of possible neighbors, j_i^k, which includes neighboring solutions that have not been visited by the kth ant. The ants' decision-making process is influenced by two non-negative parameters, α and β, which indicate the relative importance of the pheromone level (τ_i) and heuristic information (η_j), respectively.

After an ant selects the subsequent feature in its path, the newly chosen set of features is assessed using a fitness function. If the fitness value does not improve after adding any new solution, the movement of the kth ant is stopped. However, if the termination criteria are not met, the pheromone level at the i th solution for the next generation $(g + 1)$ is updated according to the information provided in reference [56].

$$\tau_i(g+1) = (1-\rho)\tau_i(g) + \sum_{k=1}^{N} \Delta\tau_i^k(g)$$
$$\Delta\tau_i^k(g) = \begin{cases} \frac{fitness(S^k(g))}{|S^k(g)|}, & if i_{\in} S^k(g) \\ 0 & otherwise \end{cases} \quad (7)$$

The pheromone decay rate ($0 \leq p \leq 1$), the number of ants (N), the quantity $|S^k(g)|$ representing the number of selected variables/features, and $\Delta\tau_i^k$ signifying pheromone deposited by the k th ant is nonzero only when the i th feature is included in the ant's shortest path; otherwise, it remains at 0.

The termination criteria for the optimization process are met when the generation count (g) reaches the predefined maximum value T. At this point, the set of solutions with the highest level of pheromone and the smallest fitness value is selected as the optimal one. A visual representation of the ACO process is shown in Fig. 2.

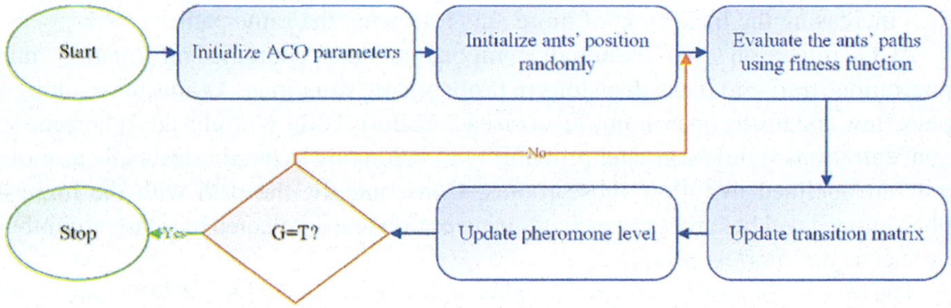

Fig. 2 Flow diagram of the ACO

3.3 Antlion-Facing Ant Colony Algorithm

The antlion optimizer (ALO) is a multistep process that includes ants' random walk, the construction of antlion traps, falling into traps, catching ants, and subsequent trap rebuilding. In ALO, the ants' walking behavior involves continuous changes in values within the source space. However, ALO is not well suited for handling some problems due to its combinatorial nature, which makes it challenging to produce effective solutions using continuous algorithms. In contrast, ant colony optimization (ACO), a discrete method, excels at addressing problems with discrete variables. In ACO, ants systematically choose edges one-by-one to discover suboptimal solutions or paths, which differs from the ants' behavior in ALO.

To overcome the limitations of ALO, a novel approach called the antlion-facing ant colony optimizer (AFACO) is proposed. AFACO combines the clever predation process from ALO with the efficient searching capability of ants in ACO. The pseudocode for AFACO is provided in Algorithm 2. One notable difference between ACO and AFACO lies in the way they record solutions. While ACO stores individual solutions, AFACO utilizes a pheromone matrix to represent and record solutions.

The main loop of AFACO is centered around the ant's search process. During this phase, an ant utilizes the chosen pheromone matrix to calculate transition probabilities and subsequently updates the pheromone matrix of the selected antlion considering its fitness. Additionally, AFACO incorporates a mutation step to further refine solutions, similar to the mutation operator employed in genetic algorithms. This mutation step aims to improve the solutions when the current ones are not optimal.

The formulation of the new pheromone matrix is defined by Eq. (8). This new approach, AFACO, combines the strengths of ALO and ACO to enhance the optimization process, making it more effective and efficient for feature selection problems. By introducing the concept of pheromone matrices and adapting the ant's movement behavior from ACO, AFACO offers a promising solution to overcome the challenges faced by ALO in tackling combinatorial optimization issues. The integration of mutation also ensures a higher likelihood of finding high-quality solutions during the search process.

$$phma_n = \omega_c \cdot phma_c + \omega_e \cdot phma_e \tag{8}$$

Here, $phma_n$ represents a new pheromone matrix. Additionally, it utilizes $phma_c$ and $phma_e$ to refer to the pheromone matrices of the chosen and elite antlion, respectively. The corresponding weights, ω_c and ω_e, are both assigned a value of 1 in this research.

The transition probability is expressed using Eq. (9).

$$P_i^k = \begin{cases} \frac{[\tau_i]^\alpha . [\eta_i]^\beta}{\sum f_{j \notin visit_k} [\tau_i]^\alpha . [\eta_i]^\beta} & f_{j \notin visit_k} \\ 0 & otherwise \end{cases} \tag{9}$$

In this context, P_i^k denotes the probability of ant k choosing feature f_i. The value τ_i corresponds to the current iteration's pheromone level of feature f_i, while η_i represents the statistical expectation heuristic information associated with selecting that feature. The set $visit_k$ encompasses the solutions that have been visited by ant k. The constants α and β are employed to control the relative importance of pheromone and heuristic information.

The pheromone matrix of the selected antlion can be updated using Eq. (10).

$$\tau_i(t) = \begin{cases} (1-\rho)\tau_i(t-1) + F(path^t)/Q & f_{i \in path^t} \\ (1-\rho)\tau_i(t-1) & otherwise \end{cases} \tag{10}$$

The evaporation rate, denoted by ρ, represents the rate at which pheromone values diminish over iterations. The route designated as $path^t$ corresponds to the better route discovered at the current iteration. The fitness value of this route is represented by $F(path^t)$, and Q is a scaling factor used in the pheromone update.

For the antlion-facing ant colony optimization (AFACO) algorithm, the parameters are configured as follows: a population size of 40 ants is used, and the maximum iterations are set to 500. The values of the parameters are specified as $\rho = 0.05$ for the evaporation rate, $\alpha = 1$, and $\beta = 2$ to control the influence of pheromone and heuristic information. The scaling factor Q is set to 0.1. Moreover, the initial pheromone value is set at 100, and the mutation probability is established as 0.1.

Algorithm 2 Pseudo-Code of AFACO

```
Input: Heuristic information, max iterations M, Population size N.
Output: Optimum (minimum) value of the function.
  Initialize antlions, compute their fitness, and update their pheromone matrices;
  Choose the elite antlion considering the best fitness;
  While termination criterion is not meet
    For every ant
      Using roulette wheel, choose the antlion's pheromone matrix;
      Combine the pheromone matrix with those of the elite one;
      A path (solution) is discovered by ant using the combination of heuristic
      information and pheromone matrix;
      Compute the fitness of the new path;
      Update the pheromone matrix of the chosen antlion considering the fitness;
    End For
    If current solutions are not superior to elite one
      Select the top one-third of antlions for mutation implementation and update
      their respective pheromone matrices;
    End If
    Update the elite solution;
  End While
```

4 Definition of the Identification Problem: Modified Bouc-Wen Model for MR Dampers

Effectively controlling and accurately accounting for the hysteretic and nonlinear behavior of MR (magnetorheological) dampers requires a model that is both manageable and highly accurate. Over time, several parametric mechanical models have been introduced to characterize the nonlinear characteristics of MR dampers [23, 30, 32]. Among these models, the extensively employed model for predicting the behavior of semi-active MR dampers is the modified Bouc-Wen model, as shown in Fig. 3. In this model, the damper force (F) is determined by Eq. (11):

$$F = \delta z + k_0(x - y) + c_0(\dot{x} - \dot{y}) + k_1(x - x_0) = c_1\dot{y} + k_1(x - x_0) \tag{11}$$

In this case, the amount of hysteretic displacement (z) can be expressed as

$$\dot{z} = -\gamma|\dot{x} - \dot{y}|z|z|^{n\prime-1} - \rho(\dot{x} - \dot{y})|z|^{n\prime} + A(\dot{x} - \dot{y}) \tag{12}$$

According to Fig. 1, $\dot{y}$ is defined by the following equation:

$$\dot{y} = \frac{1}{(c_0 + c_1)}\{\delta z + c_0\dot{x} + k_0(x - y)\} \tag{13}$$

The parameters A, ρ, γ and $n\prime$ are components of the Bouc-Wen hysteresis equation. Additionally, the parameters k_0, k_1, c_0, c_1 and δ belong to the Bouc-Wen model elements. To test the accuracy of the model under fluctuating magnetic fields, the values of δ, c_0 and c_1 in Eqs. (11) and (13) are defined as a linear function of the effective voltage u, as shown in Eqs. (14)–(16):

$$\delta(u) = \delta_a + \delta_b u \tag{14}$$

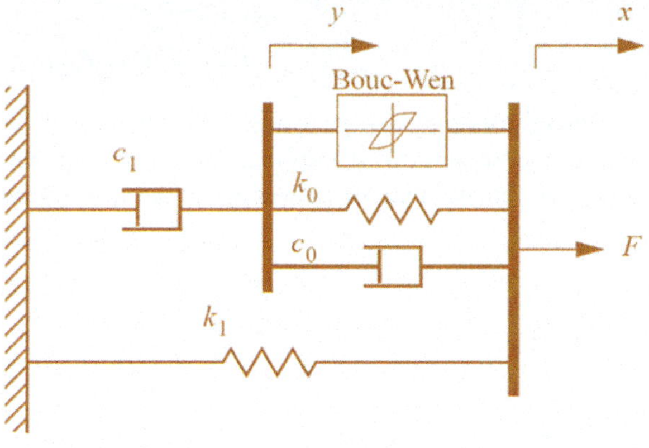

Fig. 3 The MR damper in the modified Bouc-Wen model

$$c_0(u) = c_{0a} + c_{0b}u \tag{15}$$

$$c_1(u) = c_{1a} + c_{1b}u \tag{16}$$

Taking into consideration the dynamics associated with the process of the MR fluid attaining rheological equilibrium, the effective voltage (u) is determined using a first-order filter:

$$\dot{u} = -\eta(u - v) \tag{17}$$

The relationship between the applied voltage (v) and the effective voltage (u) is expressed by Eq. (17).

In the context of this study, finding the optimal parameter values for the mathematical model is treated as an optimization problem, necessitating the use of an appropriate objective function. In this research, the objective function utilized for optimization is the normalized MSE. MSE compares the predicted response, denoted as $\tilde{f}(t_i|\boldsymbol{p})$, with the experimentally obtained response, represented by $f(t_i)$), at each time step (t_i) for a given set of parameter values (p). The mathematical expression of the objective function is given by Eq. (18).

$$OF(\boldsymbol{p}) = \frac{\sum_{i=1}^{N}\left(f(t_i) - \tilde{f}(t_i|\boldsymbol{p})\right)^2}{N.\sigma_f^2} \tag{18}$$

In Eq. (18), the variable $\boldsymbol{p}$ represents the vector containing the model parameters. σ_f^2 denotes the variance of the experimental response time history. The symbol $\sum$ denotes the summation of subsequent terms, comprising N discrete values. N corresponds to the number of experimental data points utilized in the optimization process. It is crucial to highlight that the problem seeks to optimize (minimize) the objective function while considering constraints on the parameter vector, which are described as follows:

$$\boldsymbol{p}_{min} \leq \boldsymbol{p} \leq \boldsymbol{p}_{max} \tag{19}$$

The vectors $\boldsymbol{p}_{min}$ and $\boldsymbol{p}_{max}$ are used to specify the bounds of the model parameters. These bounds are taken into consideration as constraints during the optimization process to ensure that the values of the parameters fall within acceptable ranges.

5 Numerical Study

Unknown parameters of the model must be identified for the modified Bouc-Wen model to determine the damping force for any arbitrary input. The Bouc-Wen model parameters are identified by comparing the system's realistic response and the model's numerical response in the form of an optimization problem. In this context, the identification problem of the modified Bouc-Wen model is addressed by employing the proposed optimization algorithms. The parameters k_1, k_0, c_{0a}, c_{0b}, δ_a, δ_b, c_{1a}, c_{1b}, τ, ω, A, η, and n' are selected as design variables. The enhanced and standard algorithms utilize Eq. 18 as an objective function to determine the parameters of the Bouc-Wen model. The optimization problem's constraints are considered domain limits for design variables as 50%, 100%, and 200% of realistic parameters.

To solve the identification problem, a data set of the realistic responses of the identifying system is needed. In this regard, a data set is collected from realistic responses of a 1000 kN MR damper installed in an 11-story shear building. The realistic parameters of the MR damper considering the modified Bouc-Wen model are tabulated in Table 1. The numerically generated experimental data are determined for a maximum input voltage of 10 V under the El Centro ground motion record. Briefly, experimental data acquired by modeling the 11-story shear building used for the modified Bouc-Wen model identification problem are presented in Fig. 3.

To predict the damper force, experimental data were gathered by simulating the $11 - story$ structure featuring with three 1000 kN MR dampers placed within the first three stories. The realistic parameter values obtained from Table 1 were used in these simulations. The input voltage applied to the MR dampers was generated

Table 1 Parameters of Bouc-Wen models for a 1000 *kN* MR damper

Parameter	Unit	Value
x_0	M	–
k_1	kN/m	0.0097
k_0	kN/m	0.002
c_{0a}	$kN.s/m$	110
c_{0b}	$kN.s/m/V$	114.3
δ_a	kN/m	46.2
δ_b	$kN/m/V$	41.2
c_{1a}	$kN.s/m$	8359.2
c_{1b}	$kN.s/m/V$	7482.9
ρ	m^{-2}	164
γ	m^{-2}	164
A	–	1107.2
η	s^{-1}	100
$n\prime$	–	2

through a fuzzy logic controller (FLC) and set at a maximum of 10 V. The force and displacement of the MR damper were computed numerically under the influence of the N–S component of the El Centro earthquake. These values are depicted in Fig. 4.

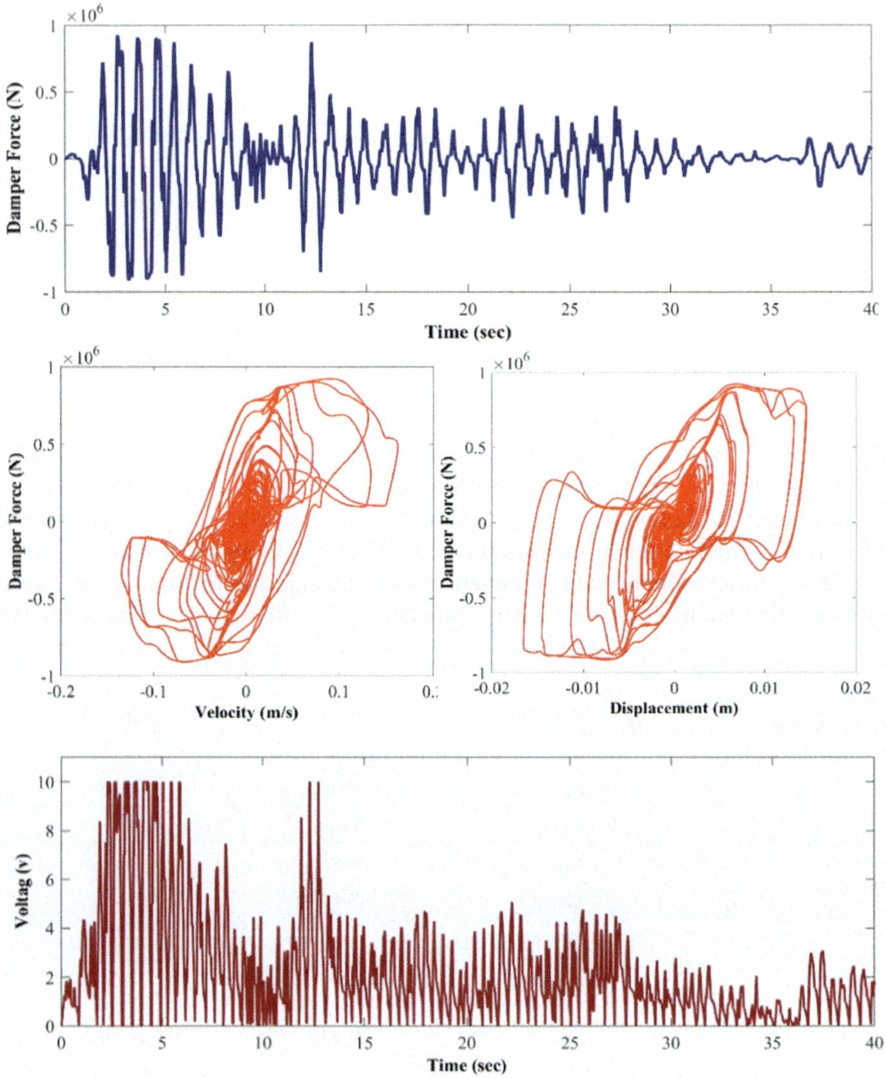

Fig. 4 Modified Bouc-Wen model under the FLC simulation

6 Results and Discussion

In this numerical example involving nonlinear systems, the proposed AFA-CO algorithm, as well as the standard ACO and ALO algorithms, are employed to determine the optimal parameter values of the MR damper with a modified Bouc-Wen model. This analysis is carried out in the context of using a fuzzy logic controller. The purpose is to compare the performance of the three algorithms and identify the most suitable parameter values for the system under consideration. The convergence history of the MSE for the standard and hybrid versions is shown in Fig. 5 on a logarithmic scale for a large domain of variables. Additionally, for the hybrid version, different search domain problems are compared in Fig. 5.

Table 2 presents the values obtained for the 13 parameters using all considered algorithms, along with the corresponding results of the MSE measure. To comprehensively evaluate the performance of the algorithms, the mentioned statistical test was conducted for the solutions of the selected algorithms. This statistical test was performed for all three domains of 50, 100, and 200% of the search space. According to the results of this test, in Table 2, the AFACO algorithm shows significantly superior performance compared to the standard versions of the algorithms. For a brief description, in the case of the 50% domain, the AFACO algorithm has relatively 73.7% and 61.3% MSE less than the ACO and ALO algorithms. Additionally, for the case of 100% domain, the EBOA algorithm has relatively 78.2% and 60.5% MSE less than the ACO and ALO algorithms. Furthermore, for the 200% domain case, the AFACO algorithm has relatively 84.5% and 79.9% MSE less than the ACO and ALO algorithms. According to the obtained results of the objective function, it is observed that the AFACO maintains its ability to solve large domain optimization problems.

To assess the accuracy and effectiveness of the AFACO in capturing the true characteristics of the damper's behavior through parameter identification for the modified Bouc-Wen model, the achieved results for the 13 parameters can be compared with their realistic values. Finally, the average error of the identified parameters compared for all algorithms in Tables 3, 4 and 5 indicate the precision of the proposed AFACO algorithm in the estimation of system parameters.

The results in Tables 3, 4 and 5 show the enhanced method's advantage in solving large domain optimization problems. Therefore, the results of system identification for a modified Bouc-Wen model of the MR damper for a 200% search domain are presented in the following. The time history of the applied damper force is compared in Fig. 6 for the experimental data and identified system using AFACO in the 200% large domain case. According to Fig. 6, the proposed AFACO exhibits a reduced error in predicting the response of a damper.

Additionally, the nonlinear hysteresis force—velocity and force—displacement curves are evaluated in Fig. 7 for the experimental data and identified systems.

Finally, the errors for the predicted forces of the MR damper model are assessed to show the accuracy of the identified system. The time history of the error and the

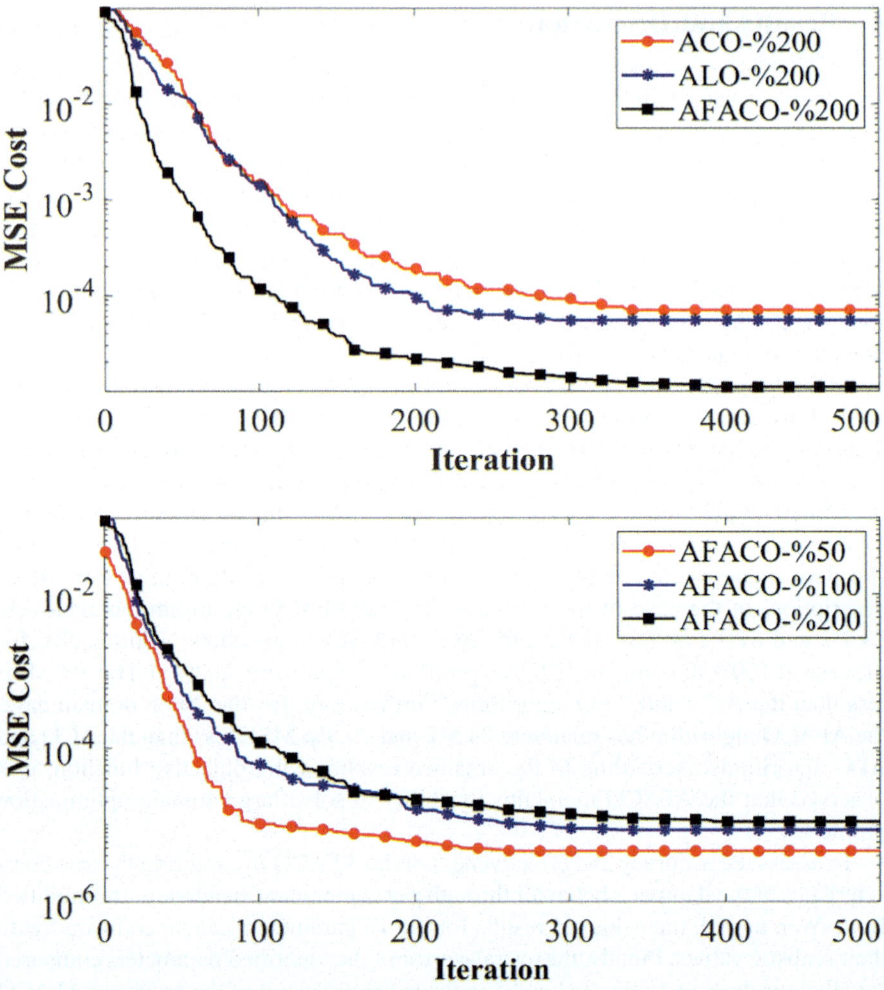

Fig. 5 Comparison of MSE convergence history for large domain case

distribution of the error are depicted in Fig. 8. To display the AFACO performance, a Gaussian normal distribution was fitted to the error distributions.

Figure 9 demonstrates that the system identified using the AFACO with a standard deviation of 1.059 is more appropriate than a system identified with the standard versions of ACO and ALO with standard deviations of 2.302 and 1.948 *kN, respectively.*

Table 2 MSE cost for various search domains using different algorithms

Domain limits		ACO	ALO	AFACO
% 50	Best	1.7713E-05	1.2025E-05	4.6512E-06
	Mean	2.6049E-04	1.6504E-04	3.3748E-05
	STD	1.0327E-04	6.7543E-05	8.5911E-06
	C_{best}	382	365	265
% 100	Best	3.9513E-05	2.1845E-05	8.6304E-06
	Mean	3.3789E-04	2.8842E-04	6.7931E-05
	STD	3.2567E-04	2.2869E-04	1.5717E-05
	C_{best}	397	425	331
% 200	Best	7.0863E-05	5.4722E-05	1.0994E-05
	Mean	2.8937E-04	1.2874E-04	5.0965E-05
	STD	1.5072E-04	1.0951E-04	4.3271E-05
	C_{best}	343	288	379

Table 3 Identified parameters set for modified Bouc-Wen with %50 domain

Parameter	ACO		ALO		AFACO	
	Value	Err. (%)	Value	Err. (%)	Value	Err. (%)
k_1	0.0098051	1.08	0.0097842	0.87	0.0094044	3.05
k_0	0.0017925	10.37	0.001847	7.65	0.0020007	0.04
c_{0a}	104.45	5.04	105.97	3.66	111.49	1.35
c_{0b}	124.33	8.77	121.95	6.7	114.50	0.17
α_a	40.186	13.02	41.814	9.49	45.994	0.45
α_b	40.735	1.13	40.866	0.81	41.383	0.44
c_{1a}	7387.07	11.63	7641.26	8.59	8435.3	0.91
c_{1b}	7719.13	3.16	7661.49	2.39	7481.20	0.02
τ	181.45	10.64	176.8	7.8	164.30	0.18
ω	166.1	1.28	165.5	0.91	165.81	1.10
A	1149.27	3.80	1139.32	2.90	1104.2	0.27
η	108.31	8.31	106.29	6.29	98.619	1.38
$n\prime$	1.9116	4.42	1.9347	3.26	1.9974	0.13
Ave. Err %		6.359		4.718		0.731
MSE	1.7713E-05		1.2025E-05		4.6512E-06	

7 Conclusions

A hybrid algorithm called AFACO was proposed in this chapter for solving the parameter identification problem in highly nonlinear systems. AFACO combines the exceptional search capabilities of ants in ACO with the impressive hunting skills of

Table 4 Achieved parameter values for modified Bouc-Wen model with 100% domain

Parameter	ACO		ALO		AFACO	
	Value	Err. (%)	Value	Err. (%)	Value	Err. (%)
k_1	0.0057286	40.94	0.0065746	32.22	0.0096601	0.41
k_0	0.0025715	28.58	0.0024497	22.49	0.0020802	4.01
c_{0a}	144.89	31.72	137.46	24.96	107.95	1.86
c_{0b}	105.99	7.26	107.84	5.65	113.10	1.05
α_a	59.851	29.55	56.983	23.34	45.937	0.57
α_b	36.083	12.42	37.139	9.86	41.475	0.67
c_{1a}	5769.71	30.98	6317.12	24.43	8581.8	2.66
c_{1b}	8049.362	7.57	7941.12	6.12	7420.50	0.83
τ	162.96	0.64	163.28	0.44	166.28	1.39
ω	186.77	13.89	181.87	10.89	163.26	0.45
A	1216.47	9.87	1193.46	7.79	1104.1	0.28
η	111.32	11.32	108.95	8.95	98.626	1.37
$n\prime$	1.7798	11.01	1.8258	8.71	2.0047	0.23
Ave. Err %	–	18.133	-	14.297		1.215
MSE	3.9513E-05		2.1845E-05		8.6304E-06	

Table 5 Achieved parameter values for modified Bouc-Wen model with %200 domain

Parameter	ACO		ALO		AFACO	
	Value	Err. (%)	Value	Err. (%)	Value	Err. (%)
k_1	0.004590	52.69	0.006011	38.03	0.009579	1.24
k_0	0.001223	38.83	0.001440	27.99	0.002124	6.21
c_{0a}	72.271	34.30	82.814	24.71	117.79	7.08
c_{0b}	136.92	19.79	130.59	14.25	119.09	4.19
α_a	41.25	10.71	42.66	7.66	44.622	3.41
α_b	45.73	11.01	44.54	8.10	41.562	0.88
c_{1a}	7823.35	6.41	7973.92	4.61	9789.1	7.11
c_{1b}	8479.1	13.31	8211.81	9.74	7340.50	1.90
τ	171.01	4.28	169.07	3.09	178.19	8.65
ω	224.38	36.82	207.37	26.44	164.06	0.04
A	1013.57	8.45	1038.97	6.16	1110.6	0.31
η	124.32	24.32	117.55	17.55	106.49	6.49
$n\prime$	1.8405	7.97	1.8858	5.71	1.9603	1.98
Ave. Err %		20.683		14.928		3.807
MSE	7.0863E-05		5.4722E-05		1.0994E-05	

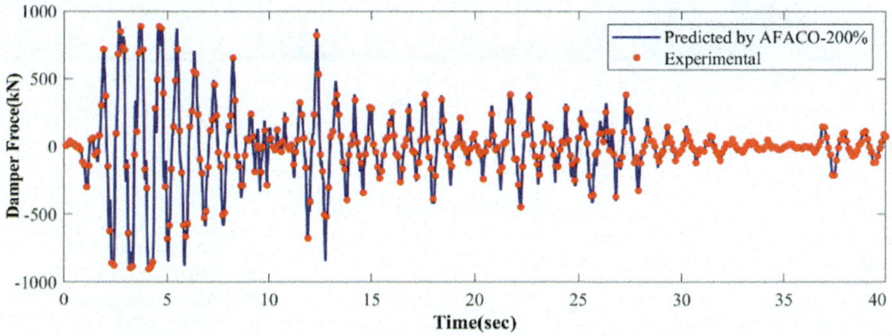

Fig. 6 Assessment of Modified Bouc-Wen Model: Experimental vs. Predicted Damper Force Responses

antlions in ALO. Moreover, AFACO incorporates a mutation operator to enhance its performance in overcoming local optima. The effectiveness and efficiency of these methods are evaluated by applying them to a parameter identification problem involving the modified Bouc-Wen model of the MR damper. The modified Bouc-Wen model consists of various parameters, each playing a crucial role in determining the damper's responses. The optimized model successfully and accurately tracks the damper's behavior compared to the outcomes achieved by using ALO and ACO alone. The findings indicate that the proposed methods exhibit excellent

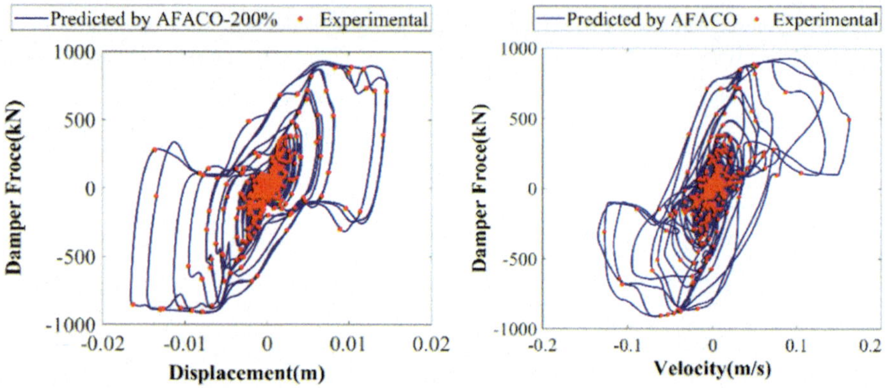

Fig. 7 Hysteresis behavior of damper for AFACO with 200% domain limit

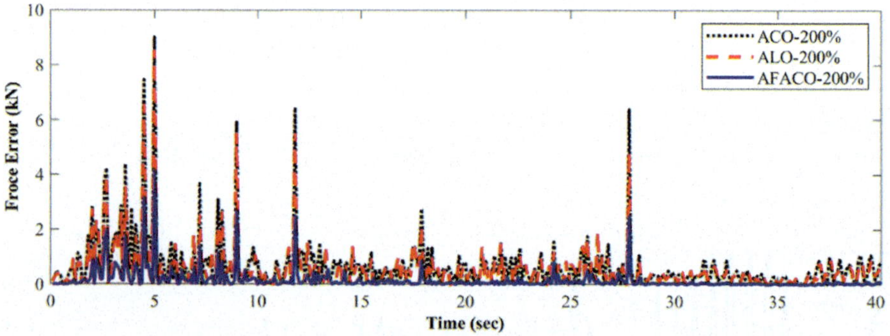

Fig. 8 Time history of damper force error for 200% domain

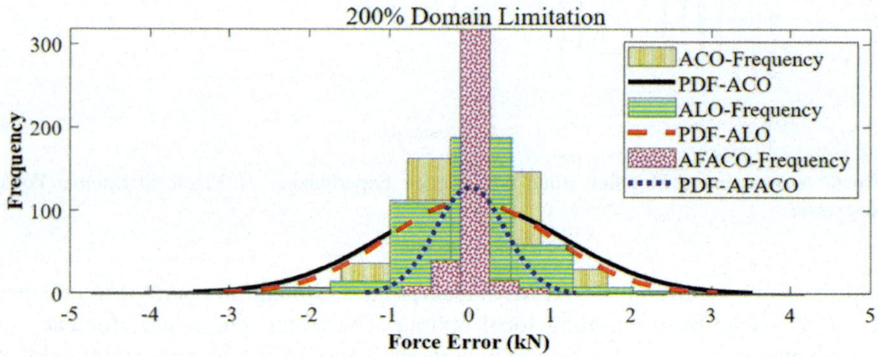

Fig. 9 Distribution of damper force error for 200% domain

capacity in addressing nonlinear issues compared to their conventional counterparts. Hybridizing the efficient features of ALO and ACO to solve the parameter identification of nonlinear systems causes an average reduction of 58.2% in the objective function compared to the standard versions. Additionally, this approach can cause an average error of 6.04% in parameter estimation in comparison to the value of 13.39% achieved by standard versions.

References

1. Dey N, Ashour A, Bhattacharyya S (2020) Applied nature-inspired computing: Algorithms and case studies. Springer, Singapore
2. Dey N (2018) Advancements in applied metaheuristic computing. Engineering Science Reference, IGI Global; Hershey, PA, USA
3. Farahmand-Tabar S, Abdollahi F, Fatemi M (2023) Robust conjugate gradient methods for non-smooth convex optimization and image processing problems. In: Kulkarni AJ, Gandomi AH (eds) Handbook of formal optimization. Springer, Singapore. https://doi.org/10.1007/978-981-19-8851-6_42-1
4. Farahmand-Tabar S, Ashtari P (2023) Bilinear Fuzzy Genetic algorithm and its application on the optimum design of steel structures with semi-rigid connections. In: Kulkarni AJ, Gandomi AH (eds) Handbook of formal optimization. Springer, Singapore
5. Farahmand-Tabar S, Ashtari P (2020) Simultaneous size and topology optimization of 3D outrigger-braced tall buildings with inclined belt truss using genetic algorithm. Struct Design Tall Spec Build 29(13):e1776. https://doi.org/10.1002/tal.1776
6. Ashtari P, Karami R, Farahmand-Tabar S (2021) Optimum geometrical pattern and design of real-size diagrid structures using accelerated fuzzy-genetic algorithm with bilinear membership function. Appl Soft Comput 110:107646. https://doi.org/10.1016/j.asoc.2021.107646
7. Farahmand-Tabar S (2023) Genetic algorithm and accelerating fuzzification for optimum sizing and topology design of real-size tall building systems. In: Dey N (eds) Applied genetic algorithm and its variants. Springer Tracts in Nature-Inspired Computing. Springer, Singapore. https://doi.org/10.1007/978-981-99-3428-7_9
8. Farahmand-Tabar S, Shirgir S (2023) Incorporating Nelder mead simplex as an accelerating operator to improve the performance of metaheuristics in nonlinear system identification. In: Kulkarni AJ, Gandomi AH (eds) Handbook of formal optimization. Springer, Singapore. https://doi.org/10.1007/978-981-19-8851-6_39-1
9. Farahmand-Tabar S, Sadrekarimi N (2023) Overcoming constraints: the critical role of penalty functions as constraint handling methods in structural optimization. In: Kulkarni AJ, Gandomi AH (eds) Handbook of formal optimization. Springer, Singapore. https://doi.org/10.1007/978-981-19-8851-6_40-1
10. Farahmand-Tabar S, Babaei M (2023) Memory-assisted adaptive multiverse optimizer and its application in structural shape and size optimization. Soft Comput. https://doi.org/10.1007/s00500-023-08349-9
11. Farahmand-Tabar S (2023) Memory-driven metaheuristics: improving optimization performance. In: Kulkarni AJ, Gandomi AH (eds) Handbook of formal optimization. Springer, Singapore. https://doi.org/10.1007/978-981-19-8851-6_38-1
12. Farahmand-Tabar S, Rashid TA (2023) Steel plate fault detection using the fitness dependent optimizer and neural networks. In: Kulkarni AJ, Gandomi AH (eds) Handbook of formal optimization. Springer, Singapore. https://doi.org/10.1007/978-981-19-8851-6_41-1
13. Farahmand-Tabar S, Shirgir S (2023) Synergistic collaboration of motion-based metaheuristics for the strength prediction of cement-based mortar materials using TSK model. In: Kulkarni

AJ, Gandomi AH (eds) Handbook of formal optimization. Springer, Singapore. https://doi.org/10.1007/978-981-19-8851-6_43-1
14. Farahmand-Tabar S, Shirgir S (2023) Positron-Enabled atomic orbital search algorithm for improved reliability-based design optimization. In: Kulkarni AJ, Gandomi AH (eds) Handbook of formal optimization. Springer, Singapore. https://doi.org/10.1007/978-981-19-8851-6_44-1
15. Farahmand-Tabar S, Ashtari P, Babaei M (2023) Dynamic intelligence of self-organized map in the frequency-based optimum design of structures. In: Kulkarni AJ, Gandomi AH (eds) Handbook of formal optimization. Springer, Singapore. https://doi.org/10.1007/978-981-19-8851-6_45-1
16. Farahmand-Tabar S, Shirgir S (2023) Boosting the efficiency of metaheuristics through opposition-based learning in optimum locating of control systems in tall buildings. In: Kulkarni AJ, Gandomi AH (eds) Handbook of formal optimization. Springer, Singapore. https://doi.org/10.1007/978-981-19-8851-6_37-1
17. Mirjalili S (2015) The ant lion optimizer. Adv Eng Softw 83:80–98. https://doi.org/10.1016/j.advengsoft.2015.01.010
18. Dorigo M, Caro GD (1999) Ant colony optimization: a new meta-heuristic. In: Proceedings 1999 congress on evolutionary computation, CEC99, vol 2, pp 1470–1477. https://doi.org/10.1109/CEC.1999.782657
19. Bouc R (1967) Forced vibration of mechanical systems with hysteresis. In: Proceedings of the fourth conference on nonlinear oscillation, Prague, Czechoslovakia. September
20. Wen YK (1976) Method for random vibration of hysteretic systems. J Eng Mech 102(2):249–263
21. Wen YK (1980) Equivalent linearization for hysteretic systems under random excitation. J Appl Mech 47(1):150–154
22. Wen YK (1989) Methods of random vibration for inelastic structures. Appl Mech Rev 42(2):39–52
23. Spencer B Jr, Sain M, Kantor J, Montemagno C (1992) Probabilistic stability measures for controlled structures subject to real parameter uncertainties. Smart Mater Struct 1(4):294
24. Song J, Kiureghian AD (2006) Generalized Bouc-Wen model for highly asymmetric hysteresis. J Eng Mech 132(6):610–618
25. Ikhouane F, Manosa V, Rodellar J (2007) Dynamic properties of the hysteretic Bouc-Wen model. Syst Control Lett 56:197–205
26. Ismail M, Ikhouane F, Rodellar J (2009) The hysteresis Bouc-Wen model, a survey. Arch Comput Methods Eng 16:161–188
27. Rakotondrabe M (2011) Bouc-Wen modeling and inverse multiplicative structure to compensate hysteresis nonlinearity in piezoelectric actuators. IEEE Trans Autom Sci Eng 8(2):428–431
28. Azar BF, Veladi H, Talatahari S, Raeesi F (2020) Optimal design of magnetorheological damper based on tuning Bouc-Wen model parameters using hybrid algorithms. KSCE J Civ Eng 24:867–878
29. Spencer B Jr, Dyke S, Sain M, Carlson J (1997) Phenomenological model for magnetorheological dampers. J Eng Mech 123(3):230–238
30. Choi SB, Lee HS, Park YP (2001) A hysteresis model for the field-dependent damping force of a magneto-rheological damper. J Sound Vib 245(2):375–383
31. Kwok N, Ha Q, Nguyen T, Li J, Samali B (2006) A novel hysteretic model for magnetorheological fluid dampers and parameter identification using particle swarm optimization. Sens Actuators A: Phys 132(2):441–451
32. Hong S, Wereley N, Choi Y, Choi S (2008) Analytical and experimental validation of a nondimensional Bingham model for mixed-mode magneto-rheological dampers. J Sound Vib 312(3):399–417
33. Graczykowski C, Pawłowski P (2017) Exact physical model of magneto-rheological damper. Appl Math Model 47:400–424
34. Bai XX, Cai FL, Chen P (2019) Resistor-capacitor (RC) operator-based hysteresis model for magneto-rheological (MR) dampers. Mech Syst Signal Process 117:157–169

35. Raeesi F, Azar BF, Veladi H, Talatahari S (2020) An inverse TSK model of MR damper for vibration control of nonlinear structures using an improved grasshopper optimization algorithm. Structures 26:406–416
36. Back T, Schwefel HP (1993) An overview of evolutionary algorithms for parameter optimization. Evol Comput 1(1):1–23
37. Sun H, Lus H, Betti H (2013) Identification of structural models using a modified Artificial Bee Colony algorithm. Comput Struct 116:59–74
38. Charalampakis AE, Dimou CK (2010) Identification of Bouc-Wen hysteretic systems using particle swarm optimization. Comput Struct 88:1197–1205
39. Kwok NM, Ha QP, Nguye MT, Li J, Samali B (2007) Bouc-Wen model parameter identification for a MR fluid damper using computationally efficient GA. ISA Trans 46(2):167–179
40. Talatahari S, Kaveh A, Rahbari NM (2012) Parameter identification of Bouc-Wen model for MR fluid dampers using adaptive charged system search optimization. Mech Sci Technol 26(8):1–12
41. Liu P, Liu H, Teng J, Cao T (2006) Parameters identification for smart dampers based on simulated annealing and genetic algorithm. In: Proceedings of the IEEE international conference on mechatronics and automation, Henan, China, June
42. Talatahari S, Rahbari NM (2015) Enriched imperialist competitive algorithm for system identification of magneto-rheological dampers. Mech Syst Signal Process 62–63:506–516
43. Shirgir S, Azar BF, Hadidi A (2020) Reliability based simplification of Bouc-Wen model and parameter identification using a new hybrid algorithm. Structures 27:297–308
44. Shirgir S, Azar BF, Hadidi A (2020) Opposition based charged system search for parameter identification problem in a simplified Bouc-Wen model. Earthq Struct 18(4):493
45. Raeesi F, Veladi H, Azar BF, Shirgir S, Jafarpurian B (2023) Upgraded salp swarm algorithm for optimal design of semi-active MR dampers in buildings. Struct Eng Mech 86(2):197
46. Hadidi A, Azar BF, Shirgir S (2019) Reliability assessment of semi-active control of structures with MR damper. Earthq Struct 17(2):131–141
47. Mirjalili S, Jangir P, Saremi S (2017) Multiobjective ant lion optimizer: a multiobjective optimization algorithm for solving engineering problems. Appl Intell 46:79–95
48. Yao P, Wang H (2017) Dynamic Adaptive Ant Lion Optimizer applied to route planning for unmanned aerial vehicle. Soft Comput 21:5475–5488
49. Mani M, Bozorg-Haddad O, Chu X (2018) Ant Lion Optimizer (ALO) algorithm. In: Advanced optimization by nature-inspired algorithms; Springer: Singapore, pp 105–116
50. Raju M, Saikia LC, Sinha N (2016) Automatic generation control of a multiarea system using ant lion optimizer algorithm based PID plus second order derivative controller. Int J Electr Power Energy Syst 80:52–63
51. Kamboj VK, Bhadoria A, Bath SK (2017) Solution of non-convex economic load dispatch problem for small-scale power systems using ant lion optimizer. Neural Comput Appl 28:2181–2192
52. Yamany W, Tharwat A, Hassanin MF, Gaber T, Hassanien AE, Kim TH (2015) A new multilayer perceptrons trainer based on ant lion optimization algorithm. In: Proceedings of the 2015 fourth international conference on information science and industrial applications (ISI), Busan, Korea, 20–22 Sept 2015; IEEE: Piscataway, NJ, USA, pp 40–45
53. Wu Y, Gong M, Ma W, Wang S (2019) High-order graph matching based on ant colony optimization. Neurocomputing 328:97–104
54. Dorigo M, Stützle T (2019) Ant colony optimization: Overview and recent advances. In: Handbook of metaheuristics; Springer, Cham, Switzerland, pp 311–351
55. Kanan HR, Faez K, Taheri SM (2007) Feature selection using ant colony optimization (ACO): a new method and comparative study in the application of face recognition system. In: Industrial conference on data mining; Springer, Berlin/Heidelberg, Germany
56. Beer C, Hendtlass T, Montgomery J (2012) Improving exploration in ant colony optimization with antennation. In: Proceedings of the 2012 IEEE congress on evolutionary computation, Brisbane, QLD, Australia, pp 10–15

Chapter 9
Hierarchical Heterogeneous Ant Colony Optimization-Based Approach to Generate Efficient Minimal Boolean Expressions for Rekeying in Authentication of IoT Devices

N. K. Sreelaja and N. K. Sreeja

1 Introduction

In today's digital world, IoT devices help connect multiple devices to the Internet [1]. IoT devices have wide usage in a variety of applications and have become a part of people's daily lives. Applications include smart cities, which are used in sensing the environment, and Industry 4.0, which is used to monitor the industry using IoT devices [1]. An IoT network consists of a set of interconnected objects that collect and transfer data over the Internet. The resources in IoT devices are constrained, which prevents them from performing efficient cryptographic operations, leading to security issues. Additionally, storage, computation, and communication are also limited for IoT devices. Other security issues, such as authentication, privacy preservation, and data integrity, must be addressed in IoT-based deployments [2, 3]. Thus, an efficient scheme for access control is essential to safeguard IoT data from tampering and unauthorized access.

The IoT environment is dynamic in nature where the IoT devices join and leave the group. This raises a security concern of protecting IoT data from unauthorized access. This can be overcome by enforcing access control. Access control in IoT devices can be ensured using group key management (GKM) [3]. In GKM, all the IoT devices in the group and their subscribers share a group key that ensures that the encrypted data by the IoT device can be accessed only by its subscribers. The shared

N. K. Sreelaja (✉) · N. K. Sreeja
Department of Applied Mathematics and Computational Sciences, PSG College of Technology, Coimbatore, India
e-mail: sreelajank@gmail.com

N. K. Sreeja
e-mail: sreeja.n.krishnan@gmail.com

N. Dey (ed.), *Applications of Ant Colony Optimization and its Variants*, Springer Tracts in Nature-Inspired Computing, https://doi.org/10.1007/978-981-99-7227-2_9

key must be changed whenever a new member joins a group or an existing member leaves the group, thereby ensuring forward secrecy and backward secrecy [4]. This mechanism suits IoT environments by replacing the need for a trusted third entity [5].

Dammak et al. [6] stated that most of the existing Group Key Management Techniques for IoT applications to ensure access control are centered on a centralized architecture. These methods are not applicable to a dynamic IoT environment which comprises multiple groups. In a dynamic IoT environment, during a join and leave event, all users in the group are affected during the rekeying process which requires to update the shared group key. Thus, a competent rekeying approach must be introduced in order to reduce the overhead in group rekeying. To overcome the drawbacks in the existing GKM techniques and design a GKM method for access control suited to the needs of a scalable and dynamic IoT environment, Dammak et al. [6] proposed DLGKM-AC, a flexible access management protocol. This protocol secures group communication, improves the overhead in rekeying during membership changes, and decreases the load on the Key Distribution Centre. The storage overhead of the devices is (log n + 3)*SK, where n is the number of devices in the group.

Sreelaja and Pai [7, 8] proposed an ACO-based approach to attain a minimal Boolean expression. Though it reduced the overhead in storage, the literal count of the Boolean expression was not focused on, which would further reduce the overhead in key storage and communication. The storage overhead in IoT devices must be minimal due to its constrained resources. Motivated by this concept, this chapter focuses on minimizing the storage and overhead in group rekeying in a dynamic IoT environment by choosing an efficient minimal Boolean expression for key distribution. The chapter proposes a hierarchical heterogeneous ant colony optimization-based approach. A binary tree representation is chosen for modeling the IoT devices in a group. Each IoT device has an identifier. A Boolean member function for a device determines the presence of an IoT device in the device group. The advantage of this approach is that storage capacity and communication overhead are efficient when compared to the existing group rekeying approaches. The contributions of this chapter are as follows:

(1) A hierarchical heterogeneous ACO-based group rekeying method for secure key distribution appropriate for numerous groups of IoT devices during a membership change, ensuring a flexible access management system.
(2) During a change in the device membership, both security requirements and backward, forward security are ensured.
(3) Experimental results prove that the storage overhead for IoT devices and communication overhead for the Key Distribution Center is a minimum.

2 Related Work

Centralized schemes for group rekeying have a single Key Distribution Center (KDC) to create and distribute keys for encryption, whereas, in a distributed scheme, the generation of group key is in a collaborative fashion among the members of the group or by a single member. In addition, there is an increased processing and communication overhead [9] in a dynamic IoT environment whenever a new device joins a group or an existing device leaves a group. Thus, the system is divided into several subgroups in decentralized schemes that reduce the load on the KDC. In addition, the group members are tracked by a subgroup manager, which further reduces the load. The Logical Key Hierarchy (LKH) [10] and the One-way Function Tree (OFT) [11] are the two major GKM schemes. LKH and OFT schemes distribute the updated encryption keys to the devices belonging to a group, centered on a symmetric key (traffic key and encryption key) approach. The drawback of the One Way Function tree is that collusion attacks are possible, and the computational overhead of the devices is increased during group rekeying. Thus, group rekeying using a one-way function tree is not perfect in IoT devices that have constrained resources. The schemes proposed in [12, 13] provided fine-grained access control attribute-based encryption (ABE) for group rekeying. However, asymmetric cryptography is employed in ABE, which is inappropriate to run on resource-constrained IoT devices [14].

In addition, Key Management Schemes in [15, 16] employ asymmetric encryption mechanisms. The group key establishment protocol proposed by Porambage et al. [17] for multicast communication employs elliptic curve cryptography. The drawback is that the protocol does not competently achieve the rekeying process. Forward secrecy is not ensured in Group rekeying proposed by Mapoka et al. [18]. An area-based multiple GKM that ensures forward secrecy was proposed by Zhong et al. [19]. However, the drawback is that the revocation events increase the overhead, making it inappropriate for vibrant IoT environments. A lightweight symmetric key establishment based on the Kronecker product was proposed by Tsai et al. [20]. However, the key updates are not considered when the devices join or leave the device group. Thus, forward secrecy and backward secrecy are not achieved in this protocol. Furthermore, a decentralized batch-based group key was proposed by Abdmeziem et al. [21]. However, large storage and computation resources are needed, which is inappropriate for IoT devices with constrained resources.

Kung et al. [22] proposed a group rekeying method to accommodate multiple device groups. A two-tier centralized system KDC was established, in which group rekeying in each device group is achieved using Logical Key Hierarchy (LKH). However, all the devices in the device group must update the keys after each join or leave event, leading to an increased computation overhead.

Dammak et al. [6] proposed a decentralized lightweight group key management architecture for access control in the IoT environment (DLGKM-AC), a flexible access management protocol. This protocol secures group communication, improves the overhead in rekeying during membership changes, and decreases the load on the

Key Distribution Centre. Group rekeying is achieved using Logical Key Hierarchy. The storage overhead in each device is (log n + 3)*SK, where n is the number of devices in the group.

A self-organized group key management (GKM) [20] protocol includes the key-encryption key (KEK) tree. Each member in the group must compute all the confidential keys from its leaf node to the root. A unique weight based on tree depth, tree weight, and node value is maintained at each node to categorize the affiliation precedence of the member node with the head. However, the limitation is that the network will be affected due to the different rules of weight challenge.

Cheikhrouhou et al. presented ring-based secure group communication (RiSeG) [23]. It involves a resource-constrained controller. The limitation is the huge latency introduced by the scheme which could not scale with a huge organization. Xiaong et al. [24] proposed a 3-phase authentication protocol for a heterogeneous IoT environment. However, the drawback is that the security features are fewer and incur a high computational and communicational cost. Gu et al. [25] proposed an IIoT-based environment that employs a Physical Layer authentication (PLA) mechanism that increases the reliability of the network. However, the drawback is that there is no security from major attacks. Shen et al. [26] proposed BASA, a blockchain-based protocol. Although the protocol is capable of providing efficient communication, the overhead in communication cost is high. Esfahani et al. [27] proposed a lightweight 2-phase mutual authentication protocol that is proven to be secure. However, the protocol has a high storage overhead. Rezabibagha et al. [28] proposed a lightweight computationally efficient protocol. Paliwal [29] proposed a lightweight protocol that is shown to be secure against attacks. Although the execution cost is low, the communication cost is high. Shuai et al. [30] proposed a secure authentication scheme that is efficient against network attacks. However, the computational and communication overhead is high. The two-factor authentication protocol proposed by Li and Wang [31] has high computational complexity.

Mughal et al. [32] proposed the LTSSM approach using mobile service computing in the IoT. However, the number of rekeying messages is Groupsize-1. A lightweight group rekeying protocol was proposed by Lakhlef et al. [33] for communicating things. Although the storage in the nodes is O (log ($n^{1/3}$)), the rekeying messages by the key manager are high depending on the group size. Kim et al. [34] proposed a secure swarm tool kit for authentication in IoT entities. The authorization entities distribute session keys securely to the IoT devices attached to a group. However, the keys stored by the authorization entity to communicate with the individual entities is "n," where "n" is the number of entities. Thus, authentication overhead is one of the research challenges to be addressed. To address this gap, a hierarchical heterogeneous ant colony optimization-based approach is proposed to generate an efficient minimal Boolean expression for key distribution and storage.

3 Hierarchical Heterogeneous Ant Colony Optimization-Based Model of Generating an Efficient Minimal Boolean Expression for Group Rekeying of IoT Devices

This section explains the key distribution by the authorization entity to the IoT devices while an IoT device joins or leaves the group. For secure transfer of messages among the IoT entities in the group, a Key Distribution Centre distributes a minimum number of keys to achieve minimum communication cost. To perform minimum cryptographic operations, the key storage and distribution must be a minimum. To achieve this, a hierarchical heterogeneous ACO-based model for generating an efficient minimal Boolean expression for group rekeying is proposed. This approach distributes the minimum number of rekeying messages and the minimum number of keys.

3.1 *Representation of IoT Devices in the Device Group*

The process of key distribution by the Key Distribution Center (KDC) is modeled as follows. The system is modeled using a binary tree denoting the IoT devices attached to each authorization entity. The Encryption Keys (EK) are held by the intermediate nodes in the tree. Each IoT device has a path key that constitutes the Keys from its leaf node to the root. The KDC constructs the binary tree as follows:

The leaf nodes denote the IoT devices. The session key (SK) is denoted using the root node. Each IoT device is identified using a device identifier. The device identifier is represented in binary form. The KDC encrypts the session key using the keys of the respective IoT devices and distributes it to the IoT devices.

The IoT devices attached to an authorization entity are represented as {IoT_1, IoT_2, IoT_3,…,IoT_{N1}}, where $N1 = 2^n$, N1 denotes the group size, and n denotes the length of bits in the binary digit depicting the IoT device (IoT_{ID}) in the device group. The IoT_{ID} is written in binary form as an n-bit ID IoT_{ID} (IoT_t) = $K_n K_{n-1} K_{n-2} \dots K_1$, where K_i is either 0 or 1. The existence of an IoT device in the device group is identified using a member function 'M()'. An IoT device is in the group if IoT_{ID} ($K_n K_{n-1} K_{n-2} \dots K_1$) = 1 and is not in the group if IoT_{ID} ($K_n K_{n-1} K_{n-2} \dots K_1$) = 0.

A session key SK and a set of keys $k_1 k_2 \dots k_n$, where k_i takes the value k_i if $K_i = 1$ and k_i^{l} if $K_i = 0$ are possessed by every IoT device in the group. The key representation for IoT devices in a group size of 8 is shown in Fig. 1. The keys for device IoT_2 (IoT_{ID} 001) possess the keys $k_3^{\mathrm{l}} k_2^{\mathrm{l}} k_1$, in addition to the session key. A truth table representation is used to denote the IoT devices in the device group and to find a minimal Boolean expression for key distribution. Table 1 shows the truth table representation of the IoT devices in the group.

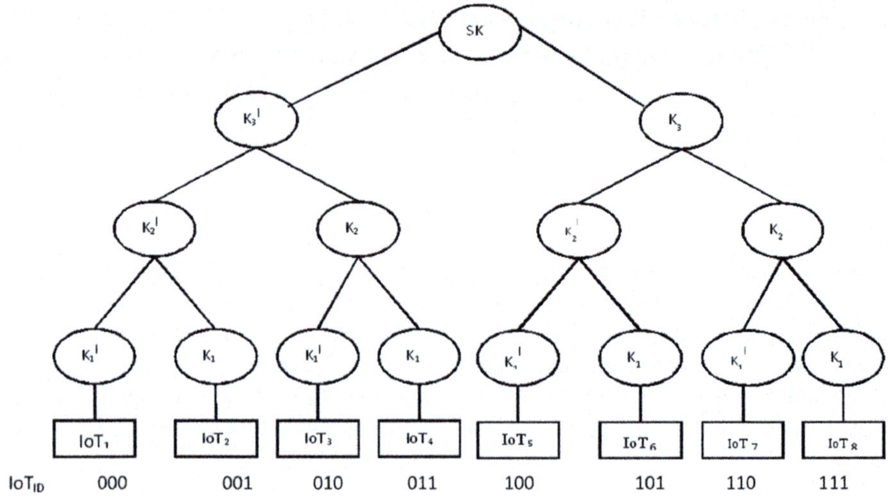

Fig. 1 Key distribution for the devices in a group

Table 1 Truth table representation of the IoT devices in a group

IoT devices	Device identifier ($K_3K_2K_1$)	Membership
IoT_1	000	1
IoT_2	001	1
IoT_3	010	1
IoT_4	011	1
IoT_5	100	1
IoT_6	101	1
IoT_7	110	1
IoT_8	111	1

3.2 HHACOBXE Algorithm for Generating Efficient Minimal Boolean Expression for Key Distribution

The IoT devices attached to an authorization entity are represented as a truth table. Every device is identified by a device identifier (IoT_{ID}). A membership value associated with the device identifier determines the presence or absence of the device in the group. The named hierarchical heterogeneous ant colony optimization Boolean expression evolver (HHACOBXE) algorithm, obtains an efficient minimal Boolean expression, based on the truth table representing the IoT devices.

The working ant agent in Level 1 computes the minimal Boolean expression for the device representation and directs to level 2 ant agent. The global value is initialized with the value 3^n-1, where 'n' is the length of bits denoting the IoT device identifier.

The level 2 ant agent counts the number of variables in the minimal Boolean expression and updates the global value. The expression that has the minimum number of variables is chosen as the solution.

A minimal Boolean expression is found by a collection of working ant agents. Initially, there are two levels. Level 2 has an ant agent and level 1 has working ant agents. Each working ant agent at level 1 deposits pheromone, which has several spots, each of which represents a term denoting a Boolean expression. Figure 2 illustrates the depiction of the pheromone deposition of the working ant agent. Each spot in the pheromone deposition denotes the terms of a Boolean expression for a case of 3 input variables. For example, the pheromone deposition {7, 8} symbolizes the terms at spots 7 and 8. Thus, an OR operator is used to combine the terms $k_1 k_2^{|}$ and $k_1^{|} k_2$ at the equivalent positions for a three-input variable case. Hence, the working ant agent's pheromone deposition {7, 8} symbolizes the Boolean expression $k_1 k_2^{|}+ k_1^{|} k_2$.

To determine if the solution is obtained, the working ant agent computes its energy by estimating the Boolean expression using the input table denoting IoT devices. The Boolean expression is estimated by substituting the input value for each literal from the truth table. The output of the estimated expression is matched with the member function M() of IoT devices to compute the energy value.

Let the member function output vector be denoted as shown in Eq. (1).

$$M = m_1, m_2, m_3, \ldots .m_N \; m_i = 0 \text{ or } 1, N = 2^n \tag{1}$$

The output vector of the working ant agent W_i is shown in Eq. (2). The energy value is calculated using Eq. (3).

SingleVariable Terms		Two Variable Terms		Three Variable Terms	
Spot	Term	Spot	Term	Spot	Term
1	k_1	7	$k_1 k_2^{|}$	19	$k_1^{|} k_2^{|} k_3^{|}$
2	$k_1^{|}$	8	$k_1^{|} k_2$	20	$k_1 k_2^{|} k_3^{|}$
3	k_2	9	$k_1 k_2$	21	$k_1^{|} k_2 k_3^{|}$
4	$k_2^{|}$	10	$k_1^{|} k_2^{|}$	22	$k_1 k_2 k_3^{|}$
5	k_3	11	$k_1 k_3^{|}$	23	$k_1^{|} k_2^{|} k_3$
6	$k_3^{|}$	12	$k_1^{|} k_3$	24	$k_1 k_2^{|} k_3$
		13	$k_1 k_3$	25	$k_1^{|} k_2 k_3$
		14	$k_1^{|} k_3^{|}$	26	$k_1 k_2 k_3$
		15	$k_2 k_3^{|}$		
		16	$k_2^{|} k_3$		
		17	$k_2 k_3$		
		18	$k_2^{|} k_3^{|}$		

Fig. 2 Representation of terms in an ant agent for a three-variable case

$$WO = wo_1, wo_2, wo_3, \ldots ..wo_N, wo_i = 0 \text{ or } 1, N = 2^n \quad (2)$$

$$E(WO_i) = count(m_i = wo_i) \quad (3)$$

The working ant agent's energy value is checked against the maximum device group size. The solution is obtained when the output value is equivalent to 2^n, and the working ant agent stops the process. All the working ant agents who have reached the solution return the minimal Boolean expression to the ant agent at level 2.

If the working ant agent does not reach the solution, the working ant agent's pheromone evaporates. To find the solution, the pheromone has to be deposited again by the working ant agent by selecting new terms. The energy for the Boolean expression depicted by the pheromone is again computed by the working ant agent. The working ant agent repeats the process until an energy value equivalent to 2^n is attained. The minimal Boolean expression is attained if any of the working ant agent reaches an energy equal 2^n, the maximum device group size. The minimal Boolean expression attained by the working ant agents is returned to the level 2 ant agent and the process is stopped.

The ant agent at level 2 counts the literals in the minimal Boolean expression. This value is compared with the global value. If the value is less than the global value, the global value is updated with the variable count value. Thus, the minimal Boolean expression with the minimum variable count is chosen as the solution.

The literals in the minimal Boolean expression represent the keys distributed to the IoT devices in the group. This KDC uses the keys of the corresponding IoT devices to encrypt the session key and distributes it to the IoT devices in the group. Each IoT device uses its keys and decrypts the session key for communication. Figure 3 illustrates the pseudocode of the HHACOBXE algorithm.

4 Case Study

This case study discusses the group rekeying operations for eight IoT devices in a group size of 8. A truth table depiction is used to show the IoT devices in a group as given in Table 2. HHACOBXE algorithm is invoked to find the minimal Boolean expression for the truth table in Table 2. The device identifier is a 3-bit binary digit represented using three variables K_3, K_2, and K_1, The binary vector from 000 to 111 signifying the IoT devices is taken as the input.

A set of working ant agents are selected to identify the minimal Boolean expression from the truth table representing the IoT devices. Every working ant agent has a collection of terms denoting the Boolean expression. The global value is 26. It is shown in Table 3 that there are four ant working agents to find the solution. The pheromone is deposited by working ant agent W_1 to find the solution. The pheromone deposition denoting the Boolean expression has only one term. Thus, the input value for the variable K_1 is passed as input to the truth table. The output is found by counting

Procedure HHACOBXE ()
K=0; global value=3^n-1;//n is the length of bits in the IoT device identifier
Choose working ant agents with terms at spots denoting Boolean expression;
Compute energy of each working ant agent according to the function
energy (WO_i)=count (m_j=wo_i);
If (energy of the working ant agent =maximum number of devices in a group) **then**
Return (boolean expression to level 2 ant agent);
Else
Repeat
Pheromone deposition of the working ant agent evaporates;
Compute the energy of the working ant agent;
If (energy of the working ant agent =maximum number of devices in a group) **then**
Return (boolean expression to level 2 ant agent);
endif;
until (energy of working ant agent= maximum number of devices in a group);
endif;
Level 2 ant agent counts the number of variables and stores in the global value;
If count(variable)< global value update global value with count;
Return (Boolean expression with minimum number of variables)
end HHACOBXE

Fig. 3 HHACOBXE algorithm pseudocode

Table 2 Truth table depiction of eight IoT devices in a group

IOT devices	Device identifier ($K_3K_2K_1$)	Output
IoT_1	000	1
IoT_2	001	1
IoT_3	010	1
IoT_4	011	1
IoT_5	100	1
IoT_6	101	1
IoT_7	110	1
IoT_8	111	1

the number of bits in the input variable matching with the bits in the member function. Thus, 4 bits match the member function, and the energy of W_1 is 4. Since the energy of W_1 is not equal to 2^n, the pheromone of W_1 evaporates and W_1 moves to the next trail. Pheromone is again deposited by W_1 and energy is found to be 8. Thus W_1 returns the minimal Boolean expression to level 2 ant agent. The process is stopped if at least 1 working ant agent reaches the solution. Thus, it is found that the solution is obtained by W_1 and W_3. The ant agent in level 2 computes the literal count in the minimal Boolean expression returned by W_1 and is found to be 2. Since this value is less than 26, value 2 is updated in the global value. Since the literal count in the minimal Boolean expression returned by W_3 is 3, the solution obtained by W_1 is chosen as the minimal Boolean expression for group rekeying. Hence, the terms at positions 1 and 2 represent the minimal Boolean expression $k_1 + k_1^{\prime}$.

The literals in terms of the minimal Boolean expression represent the keys k_1' and k_1. The keys obtained are circulated to the IoT devices in the group. A new session key *SKey(R+1)* is encrypted using the keys circulated to the IoT devices in the group. The encrypted session keys are represented as $\{SKey(R+1)\}_{k1'}$ and $\{SKey(R+1)\}_{k1}$, and communicated to the IoT devices in the group. The IoT devices IoT_1, IoT_3, IoT_5 and IoT_7 decrypts $\{SKey(R+1)\}_{k1'}$ using the key k_1'. The IoT devices IoT_2, IoT_4, IoT_6, and IoT_8 decrypts $\{SKey(R+1)\}_{k1}$, using the key k_1. Hence, the session key is stored by the IoT devices.

5 Performance Analysis of the HHACOBXE Algorithm for Efficient Rekeying with Existing Methods

The efficiency of HHACOBXE algorithm is studied in terms of storage overhead and rekeying messages with the existing approaches. Storage overhead is defined as the memory capacity required to maintain the keys.

5.1 *Comparison of the HHACOBXE Algorithm with Existing Methods in Terms of Storage Overhead*

The number of keys to be stored by the device and the Key distribution center is studied for various algorithms, such as DLGKM-AC [6], Distributed LKH [22], and HHACOBXE, as shown in Table 4. Figure 4 shows the number of keys stored in the Key Distribution Centre using HHACOBXE, DLKH [22], and DLGKM-AC [6]. It is shown that the HHACOBXE algorithm is competent in terms of key storage.

5.2 *Comparison of the HHACOBXE Algorithm and Existing Methods in Terms of Rekeying Messages*

Rekeying provides security, but it also increases computation and communication costs. Figure 5 shows a comparison of rekeying messages for a fewer number of nodes in a group. Figure 5 shows that in LNT [35], GKM [20], and RiSeG [23], more rekeying operations are needed in comparison with DBGK [21], LT-SMM [32], SST [34], and HHACOBXE. Figure 5 shows that there are more rekeying operations when the group size is higher. The number of rekeying messages in DBGK [21] and SST [34] is also approximately equal to the group size. The total number of

Table 3 Working ant agent finding minimal Boolean expression

Working Ant agent	Pheromone	Boolean expression	Output vector	Energy	Pheromone	Boolean expression	Output vector	Energy
W_1	{1}	$\{K_1\}$	{0,1,0,1,0,1,0,1}	4	{1,2}	$\{K_1 + K_1'\}$	{1,1,1,1,1,1,1,1}	8
W_2	{8,9}	$\{K_1'K_2 + K_1 K_2\}$	{0,0,1,1,0,0,1,1}	4	{9,1}	$\{K_1 K_2 + K_1\}$	{0,1,0,1,0,1,0,1}	4
W_3	{7}	$\{K_1 K_2'\}$	{0,1,0,0,0,1,0,0}	2	{3,7,4}	$\{K_2 + K_1 K_2' + K_2'\}$	{1,1,1,1,1,1,1,1}	8
W_4	{2}	$\{K_1'\}$	{1,0,1,0,1,0,1,0}	4	{4}	$\{K_2'\}$	{1,1,0,0,1,1,0,0}	4

rekeying operations in LT-SMM [32] is Group Size–1. A comparison is performed with GREP protocol [36], a Versatile Key Management protocol (VKM) [37], Light Weight Group Rekeying protocol (LWGK) [33] and SDLKH [38] with HHACOBXE for a large number of nodes as shown in Fig. 6. Thus, it is shown that HHACOBXE algorithm is efficient in terms of rekeying messages irrespective of the number of nodes. Thus, from Figs. 5 and 6, it is evident that the number of rekeying messages is minimal using the HHACOBXE algorithm.

6 Security Analysis

The effectiveness of the HHACOBXE algorithm for key distribution in IoT devices is analyzed in terms of forward and backward secrecy.

Table 4 Comparison of key storage HHACOBXE and existing methods

Algorithm	Number of keys to be stored by the device (n devices, SKey is the session key)	Number of keys to be stored by the key distribution center (n devices)
DLGKM-AC	((log n) + 3)* *SK*ey	2n−1
Distributed LKH	–	2n−1
HHACOBXE	**log n**	**log n**

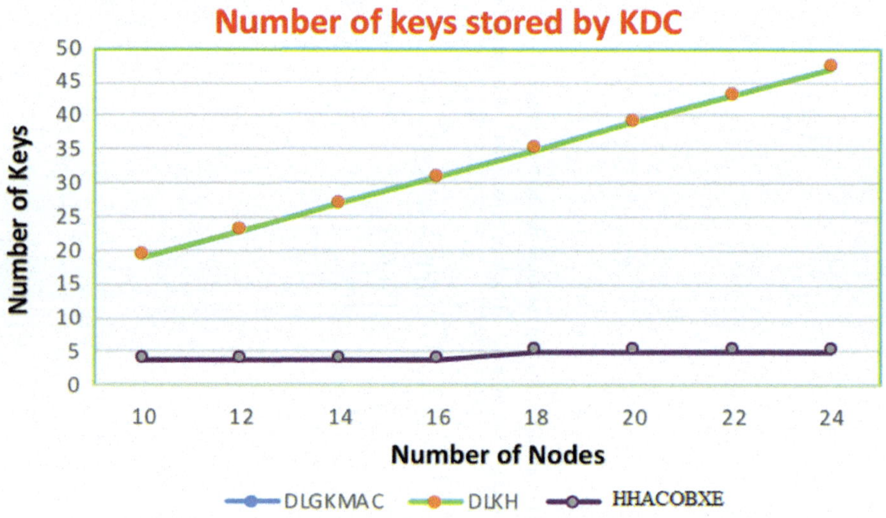

Fig. 4 Key storage in KDC using HHACOBXE and existing methods

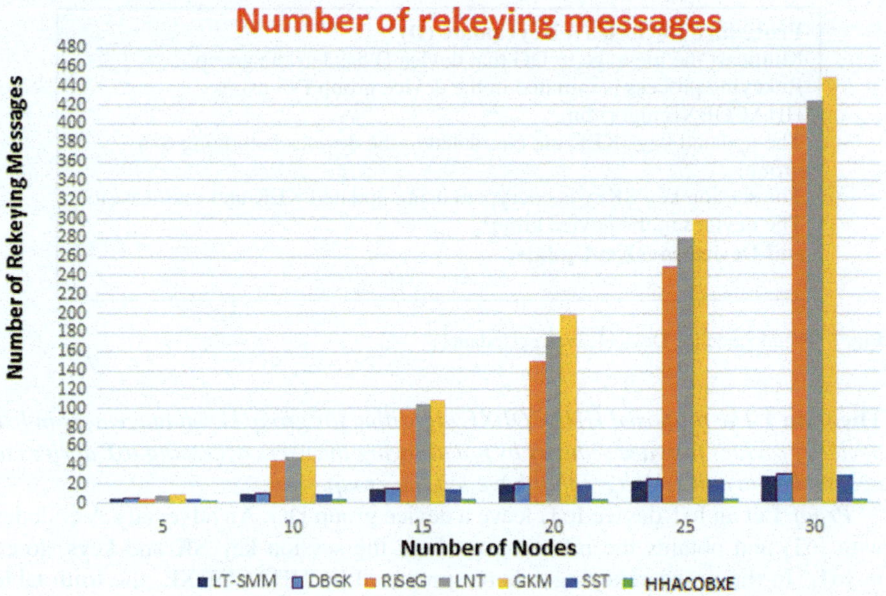

Fig. 5 Number of rekeying messages using HHACOBXE and existing methods for fewer IoT devices in a group

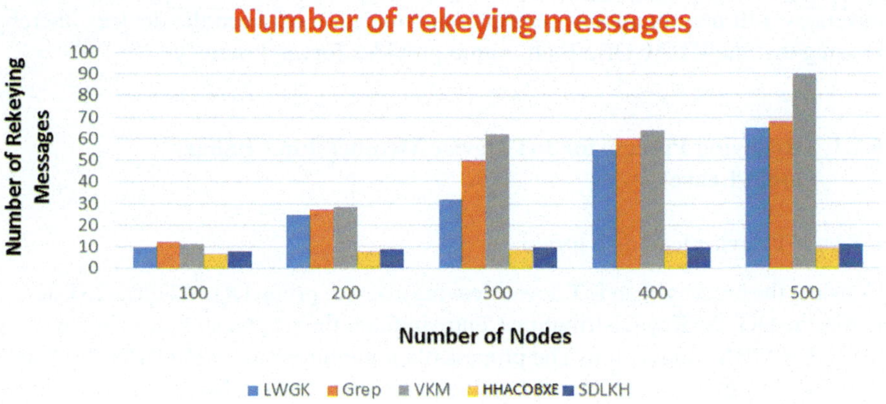

Fig. 6 Number of rekeying messages using HHACOBXE and existing methods for IoT devices in a group

6.1 Forward Secrecy

The forward secrecy of the protocol is analyzed here.

```
Procedure DeviceleaveKeyUpdate (D)
Multicast the message to DG that device D has left the group.
Rekeying process is initiated in the device group DG using
HHACOBXE algorithm.
The updated keys KEK are circulated to the devices belonging to a
group;
The session key SKey is encrypted using updated KEK and multicast to
The devices in the device group;
end DeviceleaveKeyUpdate;
```

Fig. 7 Pseudocode for DeviceleaveKeyUpdate ()

Theorem 1 *The proposed HHACOBXE algorithm for group key management in IoT devices provides forward security by not allowing the users accessing IoT devices to access the current messages that have left the group.*

Proof Let an IoT device IoT_j leave a device group DG. An adversary A colludes with IoT_j and obtains the information about the session key SK and keys stored in IoT_j. In the group rekeying scheme proposed by HHACOBXE, the truth table depicting the IoT devices belonging to a group changes during an eviction of a device IoT_j, and minimal Boolean expression is attained. Thus, the updated group key $SKey^{l} = h(SKey)$, and keys are multicast to the IoT devices in the group DG. Hence, adversary A cannot decrypt the rekeying message and obtain $SKey$. Therefore, the adversary will not obtain any current communication between the devices, thereby proving that the HHACOBXE algorithm provides forward secrecy.

6.1.1 Rekeying Process for IoT Device Membership Changes (Join/Leave)

i. An IoT device leaving a group:

Consider the process of an IoT device leaving a device group DG. When an IoT device leaves the DG, the Key Distribution Centre initiates the rekeying process by invoking the HHACOBXE algorithm. The process of employing the HHACOBXE algorithm generates a minimum number of keys to be circulated to the IoT devices belonging to a group. The session key SKey is multicast to the devices by encrypting them using the keys of the devices belonging to a group. Figure 7 shows the pseudocode for DeviceleaveKeyUpdate ().

6.2 Backward Security

An analysis of the backward security of the HHACOBXE algorithm is deliberated in this section.

```
Procedure DevicejoinKeyUpdate (D)
Add device D to Device group DG;
A shared secret key is shared with D in DG.
Rekeying process is initiated in the device group DG using
HHACOBXE algorithm.
The updated keys KEK are distributed to the devices in the group;
The session key SKey is encrypted using updated KEK and multicast
to the devices in the device group;
end DevicejoinKeyUpdate;
```

Fig. 8 Pseudocode for DevicejoinKeyUpdate

Theorem 2 HHACOBXE algorithm provides backward security by not allowing the IoT device to access the previous messages that have newly joined the group.

Proof Assume that an IoT device IoT_j joins the device group DG. In the HHACOBXE algorithm, the truth table depicting the IoT devices belonging to a group changes during an addition of a device, and minimal Boolean expression is attained. Thus, new session Key SKey and new keys are distributed to the device IoT_j. Thus, the newly added device IoT_j can neither derive the session key SKey nor acquire the preceding session keys.

ii. An IoT device joining a group:

Consider the process of an IoT device added to a device group DG. During the joining process, the Key Distribution Centre initiates the rekeying process by invoking the HHACOBXE algorithm. The process of employing the HHACOBXE algorithm generates a minimum number of keys to be circulated to the devices belonging to the group. The session key SKey is multicast to the devices by encrypting them using the keys of the devices belonging to the group. Figure 8 shows the pseudocode for DevicejoinKeyUpdate ().

7 Conclusion

An HHACOBXE algorithm is proposed to generate efficient minimal Boolean expressions for group rekeying in the authentication of IoT devices. The minimal Boolean expression represents a minimum number of keys to be stored and distributed to the devices belonging to the group. The advantage of the HHACOBXE algorithm is that the computational and storage overheads for IoT devices and servers are minimal compared to the existing group rekeying methods. Security requirements, such as backward and forward secrecy, are ensured. The algorithm is shown to be competent in terms of computational complexity. Additionally, fewer keys are stored compared to the existing approaches.

References

1. Zanella A, Bui N, Castellani A, Vangelista L, Zorzi M (2014) Internet of things for smart cities. IEEE Internet Things J 1(1):22–32
2. Dammak M, Boudia ORM, Messous MA, Senouci SM, Gransart C (2019) Token- based lightweight authentication to secure IoT networks. In: 2019, 16th IEEE annual consumer communications & networking conference (CCNC), Las Vegas, NV, USA, pp 1–4
3. Grammatikis PIR, Sarigiannidis PG, Moscholios ID (2019) Securing the internet of things: challenges, threats, and solutions. Internet of Things 5
4. AlMajed HN, AlMogren AS (2019) Simple and effective secure group communications in dynamic wireless sensor networks. Sensors, 19
5. Banks A, Gupta R (2014) MQTT version 3.1.1. OASIS standard
6. Dammak M, Senouci S-M, Messous MA, Elhdhili MH, Gransart C (2020) Decentralized lightweight group key management for dynamic access control in IoT environments. IEEE Trans Netw Serv Manag 17(3):1742–1757. hal-02965346
7. Sreelaja NK, Vijayalakshmi Pai GA (2014) Swarm intelligence based approach for sinkhole attack detection in wireless sensor networks. Appl Soft Comput 19:68–79
8. Sreelaja NK, Vijayalakshmi Pai GA (2009) Ant colony optimization based message authentication for wireless networks. In: Feng J (ed) Wireless networks: research technology and applications. Nova Science Publishers, pp 373 386
9. Karuturi NN, Gopalakrishnan R, Srinivasan R, Rangan CP (2008) Foundations of group key management-framework, security model and a generic construction. In: IACR cryptology EPrint
10. H. Harney and E. Harder, "Logical key hierarchy protocol," Internet draft, Tech. Rep., 1999.
11. Balenson D, McGrew D, Sherman A (2000) Key management for large dynamic groups: One-way function trees and amortized initialization
12. Ding W et al (2019) An extended framework of privacy-preserving computation with flexible access control. IEEE Trans Netw Serv Manag TNSM
13. Nabeel M, Shang N, Bertino E (2013) Privacy preserving policy-based content sharing in public clouds. IEEE Trans Knowl Data Eng 25(11):2602–2614
14. Wang X, Zhang J, Schooler EM, Ion M (2014) Performance evaluation of attribute- based encryption: toward data privacy in the IoT. In: IEEE ICC.
15. Sciancalepore S, Capossele A, Piro G, Boggia G, Bianchi G (2015) Key management protocol with implicit certificates for IoT systems. In: Proceedings of the 2015 workshop on IoT challenges in Mobile and Industrial Systems. ACM, pp 37–42
16. Tseng Y, Fan C, Wu C (2019) FGAC-NDN: fine-grained access control for named data networks. IEEE Trans Netw Serv Manag 16(1):143–152
17. Porambage P, Braeken A, Schmitt C, Gurtov A, Ylianttila M, Stiller B (2015) Group key establishment for enabling secure multicast communication in wireless sensor networks deployed for IoT applications. IEEE Access 3:1503–1511
18. Mapoka TT, Shepherd SJ, Abd-Alhameed RA (2015) A new multiple service key management scheme for secure wireless mobile multicast. IEEE Trans Mobile Comput 14(8):1545–1559. (1 Aug 2015)
19. Zhong H, Luo W, Cui J (2017) Multiple multicast group key management for the Internet of People. Concurr Comput Pract Exper 29:e3817. https://doi.org/10.1002/cpe.3817
20. Tsai I-C, Yu C-M, Yokota H, Kuo S-Y (2017) Key management in internet of things via Kronecker product. In: 2017 IEEE 22nd Pacific Rim international symposium on Dependable Computing (PRDC), IEEE
21. Abdmeziem MR, Tandjaoui D, Romdhani I (2015) A decentralized batch based group key management protocol for mobile internet of things (dbgk). In: 2015 IEEE international conference on computer and information technology
22. Kung Y, Hsiao H (2018) GroupIt: lightweight group key management for dynamic IoT environments. IEEE Internet Things J 5(6):5155–5165

23. Cheikhrouhou O, Koubâa A, Dini G, Abid M (2011) RiSeG: A ring based secure group communication protocol for resource-constrained wireless sensor networks. Pers Ubiquitous Comput 15(8):783–797
24. Xiong H, Wu Y, Jin C, Kumari S (2020) Efficient and privacy-preserving authentication protocol for heterogeneous systems. IIoT IEEE Internet Things J. 7(12):11713–11724. https://doi.org/10.1109/JIOT.2020.2999510
25. Gu Z, Chen H, Xu P, Li Y, Vucetic B (2020) Physical layer authentication for noncoherent massive SIMO-enabled industrial IoT communications IEEE Trans. Inf Forensics Secur 15:3722–3733. https://doi.org/10.1109/TIFS.2020.2998947
26. Shen M, Liu H, Zhu L, Xu K, Yu H, Du X, Guizani M (2020) Blockchain-assisted secure device authentication for cross-domain industrial IoT IEEE. J Sel Areas Commun 38(5):942–954. https://doi.org/10.1109/JSAC.2020.2980916
27. Esfahani A, Mantas G, Matischek R, Saghezchi FB, Rodriguez J, Bicaku A, Maksuti S, Tauber MG, Schmittner C, Bastos J (2019) A lightweight authentication mechanism for M2M communications in industrial IoT environment. IEEE Internet Things J 6(1):288–296. https://doi.org/10.1109/JIOT.2017.2737630
28. Rezaeibagha F, Mu Y, Huang X, Yang W, Huang K (2019) Fully secure lightweight certificateless signature scheme for IIoT IEEE. Access 7:144433–144443. https://doi.org/10.1109/ACCESS.2019.2944631
29. Paliwal S (2019) Hash-based conditional privacy preserving authentication and key exchange protocol suitable for industrial Internet of Things IEEE. Access 7:136073–136093. https://doi.org/10.1109/ACCESS.2019.2941701
30. Shuai M, Xiong L, Wang C, Yu N (2020) A secure authentication scheme with forward secrecy for industrial Internet of Things using Rabin cryptosystem Computer. Communication 160:215–227
31. Li W, Wang P (2019) Two-factor authentication in industrial Internet-of-Things: attacks, evaluation and new construction. Future Gener Comput Syst 101:694–708
32. Mughal MA, Shi P, Ullah A, Mahmood K, Abid M, Luo X (2019) Logical tree based secure rekeying management for smart devices groups in IoT enabled WSN (LT-SMM). IEEE Access 7
33. Lakhlef H, Bouabdallah A (2021) Efficient and lightweight group rekeying protocol for communicating things. Comput Electr Eng 91:107021
34. Hokeun Kim, Eunsuk Kang, Edward A. Lee, and David Broman. 2017. A toolkit for construction of authorization service infrastructure for the internet of things. In: Proceedings of the 2nd ACM/IEEE international conference on internet-of-things design and implementation, Pittsburgh, PA USA, April 2017 (IoTDI 2017). https://doi.org/10.1145/3054977.3054980
35. Cheikhrouhou O, Koubaa A, Dini G, Alzaid H, Abid M (2012) 'LNT: A logical neighbor tree secure group communication scheme for wireless sensor networks.' Ad Hoc Netw 10(7):1419–1444
36. Tiloca M, Dini G (2016) GREP: a group rekeying protocol based on member join history. In: IEEE symposium on computers and communication
37. Kandi M, Lakhlef H, Bouabdallah A, Challal Y (2020) A versatile key management protocol for secure group and device-to-device communication in the internet of things. J Netw Comput Appl 150
38. Zhu WT (2005) Optimizing the tree structure in secure multicast key management. IEEE Commun Lett 9(5)

Chapter 10
Application of Ant Colony Optimizer in Thermal Power System for Frequency Management

D. Boopathi, K. Jagatheesan, and Sourav Samanta

1 Introduction

In this chapter, demand refers to electric power demand. The development of any country depends on electricity. In recent days, all fields have been automated and modernized by electricity. The individual requirement for electricity also increases every day. That demand has to be met by increasing power plant capacity and installing new power generating units. Increasing the capacity of the plant leads to complexity in the power network. When the network is complex, maintenance is too difficult. Maintenance in the sense of keeping power quality within standard limits. In the power system, frequency is a major parameter to maintain at normal at all times. To maintain and control the frequency deviation from the standard values in the power network, the load frequency control (LFC) scheme is commonly implemented. Numerous researchers are dealing with the LFC with optimization methods and controllers.

ACO was utilized to enhance the PID controller performance for a microgrid, and the result of ACO was compared with PSO [1]. A hybrid multi-area power network was examined by ACO–PID for AGC, and the results of the Artificial bee colony, GA, and improved PSO methods were compared [2]. An interlinked thermal power network with PV has been examined by a PSO-tuned fuzzy controller, and the Firefly and genetic algorithm behavior was equated to show the improvement of the PSO method [3]. The author proposed a novel lightning attachment procedure optimization technique to optimize a linear quadratic Gaussian-based PID controller to debute

D. Boopathi (✉) · K. Jagatheesan
Department of EEE, Paavai Engineering College, Namakkal, Tamil Nadu, India
e-mail: boopathime@gmail.com

S. Samanta
University Institute of Technology, Burdwan, West Bengal, India

N. Dey (ed.), *Applications of Ant Colony Optimization and its Variants*, Springer Tracts in Nature-Inspired Computing, https://doi.org/10.1007/978-981-99-7227-2_10

a three-area wind farm integrated thermal power plant LFC [4]. The pheromone-improved ACO technique tuned the FPID controller adopted in interlinked thermal power system for LFC. The improvement and supremacy were found by comparing the results of the ZN, GA, and hPSO techniques [5]. A two-area coal-based power network was investigated for an LFC by an ACO–PID controller [6]. A symbiotic organism search (SOS)-based PID controller is employed as an auxiliary control for interlinked power grids. SOS is a novel technique derived by the author [7].

The chaotic FA-tuned PID controller investigated an interlinked power network using the ITAE objective function, and the enhancement of the CFA was studied by comparing the FA and enhanced FA results [8]. The CFA provides superior control over different FAs in various working scenarios. The PSO–PID controller for the LFC was used to study a standalone power system with various sources. The performance of the classical controllers was compared with suggested controller to determine which controller was better [9]. Self-adaptive multipopulational-based Jaya optimizer tune controller (PID) designed for interconnected thermal power generating network LFC [10], the effect of the nonlinear components is analyzed in detail. An interconnected thermal power plant with GRC was examined by ACO–PID controller for LFC. The dominance of ACO was validated by the Craziness-based PSO technique [11]. A hybrid FA and pattern search -based FPID regulator has been developed for two-area thermal power plants LFC, and the hFA is demonstrated by comparing the result with DE and craziness-based PSO. Additionally, the effect of GRC on thermal power plants was analyzed [12].

An isolated power system was inspected by a PSO scheme designed controller (PID), and the performance was validated with DE, GA, and classical-based controllers. The proposed PSO-PID controller provided 79%, 55%, and 24% improvements over conventional, GA, and DE-based PID controllers [13]. Modified tilted integral derivative (TID) controller tuned by Archimedes optimization algorithm for LFC of thermal, hydro, and gas sources-based power network. The dominance of the AOA–MTID controller was proved by comparing the response of the AOA-tuned I and PI controllers [14]. Wind and solar integrated thermal standalone microgrid was examined by MA–PID controller. The results of DE, PSO, and GA–PID results was compared, to demonstrated the influence of the MA-PID controller [15]. A multi-area thermal power network with GRC is investigated by a mine blast algorithm-tuned FPID controller for frequency regulation. Antlion optimizer and hPSO-based FPID controller results are utilized by the author for validating the proposed controller results. The fuzzy controller triangle membership function is used for tuning the controller gain values [16].

A new artificial intelligence algorithm, the gravitational search algorithm and bat-inspired algorithm, is developed to tune the model predictive controllers to be employed as a subordinate controller for a nonlinear interconnected thermal power network. Additionally, capacitive energy storage and superconducting magnetic energy storage are included in the system. The effect of the energy storage systems is analyzed with conventional and ZN methods [17]. An improved Rao optimization algorithm premeditated (PID) controller was appointed in hybrid power system integrated with thermal, PV, and wind sources as a secondary controller. The algorithm is designed in three different stages. From the result comparison, the Rao_3 algorithm performed well over all others [18]. The ACO technique-tuned PID controller has been developed for a pure renewable source-based standalone power system LFC. The system is constructed with a wind source as the major component, and energy storage units are incorporated. The ACO–PID controller stabilized the oscillation more than the conventional PSO method [19]. Slime mold algorithm-based PI controller designed for an interconnected thermal power network LFC. The GA and FA-PI controller results were compared to display the authority of the projected controller [20].

An interlinked power system with GRC and boiler components was investigated by a harmony search-based PID controller. The GA–PID controller result was compared. The suggested controller controlled the fluctuation of the system frequency during uncertainty of load demand [21]. Dandelion Optimizer-tuned PI–PD controller developed for a nonlinear included coal-based power system. The Archimedes optimization algorithm and earning performance-based behavior optimizer are utilized to optimize the proposed controller [22]. The impact of the various energy storage units on a nuclear power plant is analyzed in detail by the author using ACO–PID controller. The energy storage units are supported to the secondary control loop during emergency loading conditions [23]. The author in [24] proposed the bird swarm algorithm-based TIDD controller for the LFC of a multiple area power network. The network comprises thermal and dish Stirling solar thermal sources. The response of the bird swarm algorithm was compared with crow search and cuckoo search. Optimization techniques not only for LFC, they are widely used for many engineering problems like bat-inspired algorithm for power transmission planning [30], Binary grey wolf optimizer for unit commitment problems [31]. The comparative review is reported in Table 1.

The above literature review helped to find the exact study needs in thermal power systems. There are N number of researches studied the LFC in thermal power plant as a single or interconnected system. But, in this proposed chapter a detailed study is conduct for single/two/three area systems to demonstrate the heightening of the ACO technique. This chapter reviews the performance of the controller (ACO–PID) in the proposed systems. This projected study will explain the response of the ACO–PID controller in an isolated and interlinked area thermal power generating system.

1.1 Contribution of the Chapter

- Size incremental power system designed for analysis (single/two/three area).
- An auxiliary controller loop was developed for the proposed system.
- Ant colony optimization technique utilized to enhance the control action of the secondary control loop.

Table 1 Comparative review

Ref. no	Method of tuning	Controller	System	Study comparison
[1]	ACO	PID	Microgrid—Wind and Solar	PSO
[2]	ACO	PID	Hybrid multi-area	GA, ABC, and improved PSO
[3]	Lightning attachment procedure optimization	Fuzzy controller	Two area—thermal	GA and FA
[5]	Improved ACO	FPID	Two area—non-reheated thermal	ZN, GA, and hPSO
[6]	ACO	PID	Two area—non-reheated thermal	Conventional
[9]	PSO	PID	Single area—thermal, gas, and hydro	Classical controllers (P, PI, PID)
[11]	ACO	PID	Two area—thermal	Craziness PSO
[12]	hPSO	FPID	Two area—thermal + GRC	Craziness PSO and DE
[13]	PSO	PID	Single area—thermal, hydro, and gas	Conventional, GA, and DE
[14]	ACO	PID	Single area—wind, energy storage units	Conventional and PSO

- The presentation and supremacy of the projected (ACO–PID) controller were validated with PSO and GA techniques.
- The proposed ACO-based controller provides improved results over the GA and PSO techniques.

1.2 Structure of the Chapter

This chapter is formatted as follows. The outline and detailed literature review are addressed in Sect. 1. The design of the proposed power producing system and its transfer functions are accessible in Sect. 2. Furthermore, Sect. 3 explains the method of optimization and the structure of the secondary controller. The results of ACO–PID are discussed in Sect. 4. Finally, the conclusion is presented in Sect. 5.

2 System for Investigation

The proposed chapter has three power system models for investigating the response of the ACO–PID controller against frequency oscillations at the time of unexpected load demand occurrence. Thermal power plants are the most conventional method of electric power generation. This plant comprises a governor, turbine, reheater, and rotating power system. The governor controls the input to the turbine, and the turbine converts the pressure energy into mechanical (rotating) energy. A generator that is connected to a turbine produces electricity by rotating mechanical energy. In this chapter, three different models are proposed for the investigation of single, two, and three areas, as shown in Figs. 1, 2, and 3, respectively. The transfer function model of the blocks in the thermal system is given in Table 2 [6, 11].

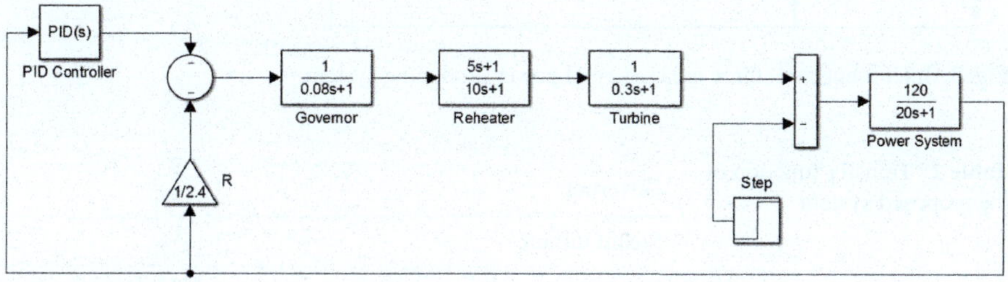

Fig. 1 Standalone thermal power system

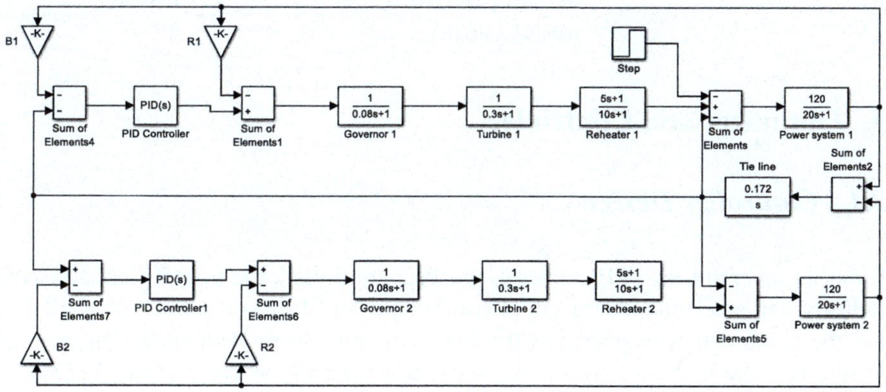

Fig. 2 Interconnected (two area) thermal power producing system

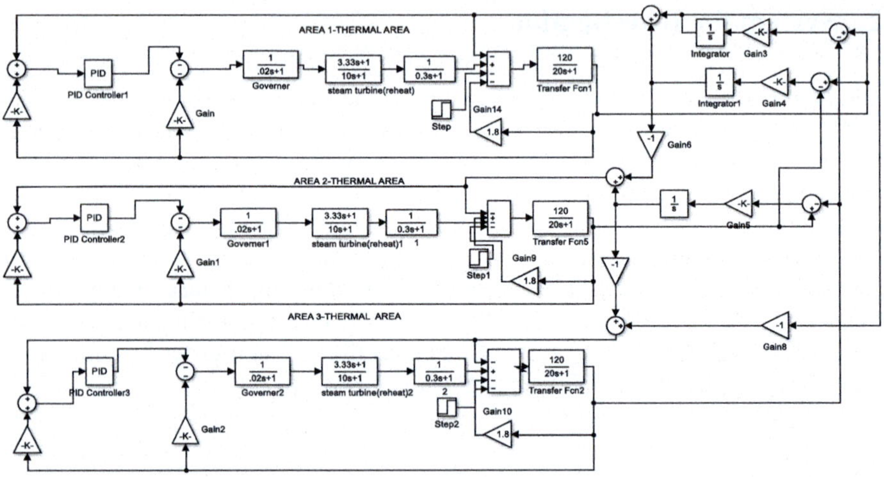

Fig. 3 Interconnected (three area) thermal power producing system

Table 2 Transfer function of the proposed system

Governor	$\frac{1}{1+sT_g}$
Steam turbine	$\frac{1}{1+sT_t}$
Reheater	$\frac{1+sK_rT_r}{1+sT_r}$
Rotating mass (Power system)	$\frac{K_p}{1+sT_p}$
Regulator gain (R)	1/R
Biasing gain (B)	B

3 Optimizer and Controller

3.1 Controller Design

From its invention in 1910 onward, the PID controller was broadly used in both industry and academia. A variety of combinations of PID controllers are available, but the combination available in the Fig is commonly implemented. The general synthesis at the PID controller is derived mathematically as given in Eq. 1 [27].

$$\text{Controlled output } U(t) = K_p e(t) + K_i \int_0^t e(t)dt + K_d \frac{de(t)}{dx} \tag{1}$$

Equation e(t) points at the error signals K_p, K_i, and K_d at the controller gain values of the proportional (P), integral (I), and derivative (D) controllers. Similarly, t represents the simulation time, and u(t) denotes the control signal produced by the

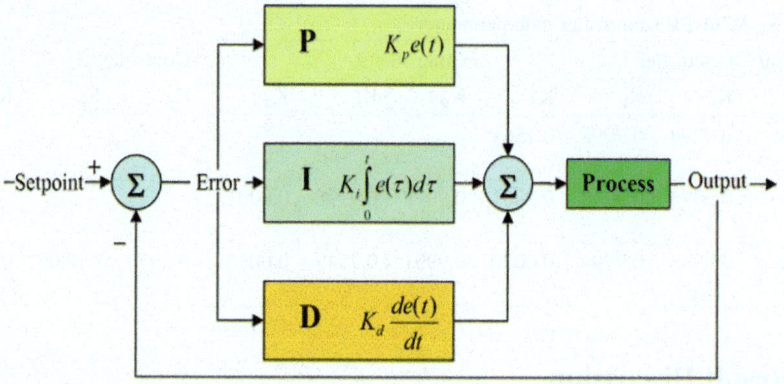

Fig. 4 The PID controller typical assembly

controller. The e(t) represents error signal, difference between the set point plant output g(t). In PID controller taking came about the steady state error in system resume, time-state controller concentrates on setting time at the system similarly. The derivative controller takes care of the stabling of the system. To achieve the above criterion, we need to select an appropriate gain value for setting the expected performance of the system. In this connection, the ACO technique is proposed to optimize the controller gain value. The details of the technique are given below. A typical assembly of the PID controller is given in Fig. 4 [28].

3.2 Ant Colony Optimization

The ACO approach was encouraged by ant colony behavior, and the procedure was created in the 1990s by Marco Dorigo. Ants in the colony have started their first random search for food around the nest. Pheromones are the way that ants in a group talk to each other about food. When the ants go back to the nest, based on how much and how good the food is, the pheromone is evaporated. All ants follow the highly concentrated pheromone trail beginning with the second excursion. Finally, ants discover the quickest way between their colony and food. Similarly, the method used objective functions to find the lowest fitness values in each iteration. ITAE (integral temporal absolute error) is used in this study. Each rehearsal has a minimum fitness value called the local best, and the final method provides a minimum value called the global best. Figure 3 depicts the ACO functional flow chart. ACO is perfectly utilized for proposed controller gain parameters tuning [25, 26, 29]. The optimized parameters are reported in Table 3.

Table 3 ACO–PID controller gain parameter

System/ Gain	Controller 1			Controller 2			Controller 3		
	K_p	K_i	K_d	K_p	K_i	K_d	K_p	K_i	K_d
Single area	0.9998	0.9998	0.1565	–	–	–	–	–	–
Two area	0.9998	0.9998	0.2801	0.9998	0.9998	0.00013	–	–	–
Three area	0.9996	0.9996	0.0020	0.9991	0.2549	0.0035	0.9979	0.6805	0.0142

4 Result Discussion

In this part, we discuss how ACO–PID works in standalone and interconnected power producing systems. In three different cases, the responses of the single-area, two-area, and three-area thermal power systems are discussed. Case 1 is about determining how the ACO–PID controller works in a power system that works on its own. Case 2 is a study of the reaction of the ACO–PID controller in two thermal power systems. Case 3 involves the reaction study of an ACO–PID controller in three-area thermal power systems. The suggested Simulink models are run through MATLAB 2016a working software with 1% SLP.

4.1 Case 1: Response Analysis of ACO–PID Controller in a Single-Area (Thermal Power System)

In this section, we discuss a full study of the results of a standalone thermal power producing system. Figure 5 shows the frequency variation (delF) comparison, and Table 4 shows the theoretical comparison. The time-domain specific metrics settling time ($T_{setlling}$), peak overshoot ($P_{overshoot}$), and peak undershoot ($P_{undershoot}$) are discussed, and the improvement is given.

In a standalone coal-based power system, the proposed ACO–PID controller controlled the frequency oscillation more quickly than the GA and PSO methods. ACO–PID settled the oscillation at 19 s < PSO: 22 s < GA:23 s. The result is plotted in a column chart for clear demonstration and shown in Figs. 6 and 7.

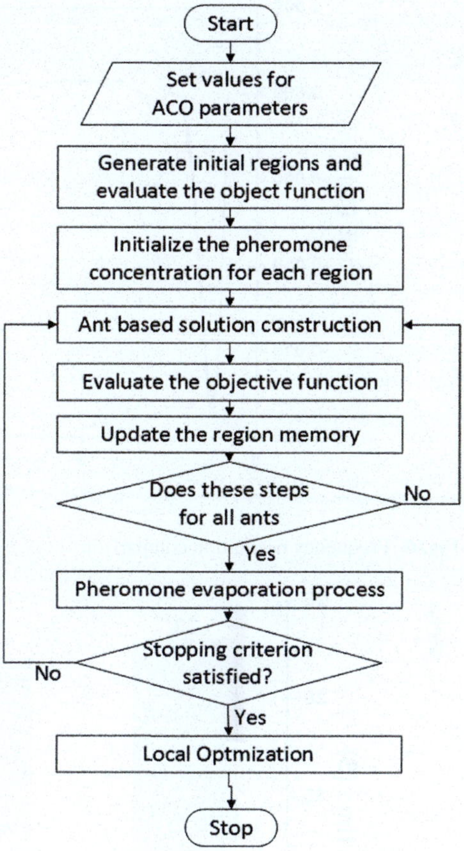

Fig. 5 ACO functional flow chart

Table 4 Controlled time-domain specific parameters in a standalone thermal power system

Time domain constraints/ Controller	Ts (S)	POs (Hz) X 10–3	PUs (Hz)
GA–PID	23	1	0.0175
PSO–PID	22	1	0.0185
ACO–PID	19	0.9	0.0175

4.2 Case 2: Response Analysis of ACO–PID Controller in an Interconnected Thermal System

The interconnected coal-based power system is investigated for LFC using ACO–PID controller. The system delF (area 1 and 2 and the tie-line power flow response) comparison are shown in Figs. 8, 9, and 10, respectively. The numerical values in the response comparison are reported in Table 5.

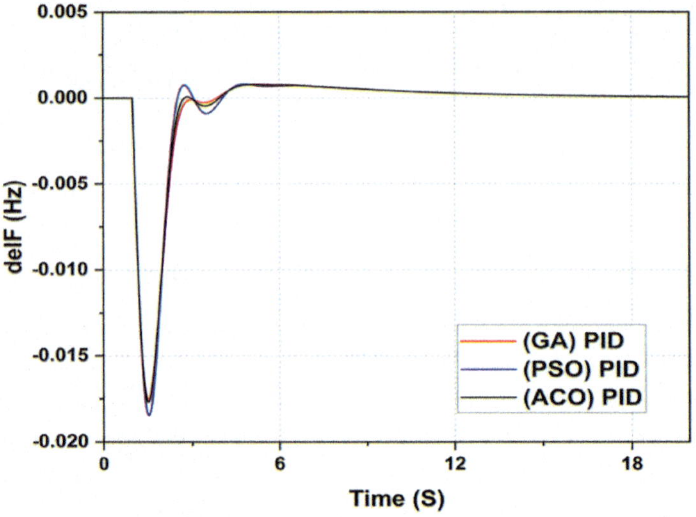

Fig. 6 Frequency oscillation comparison

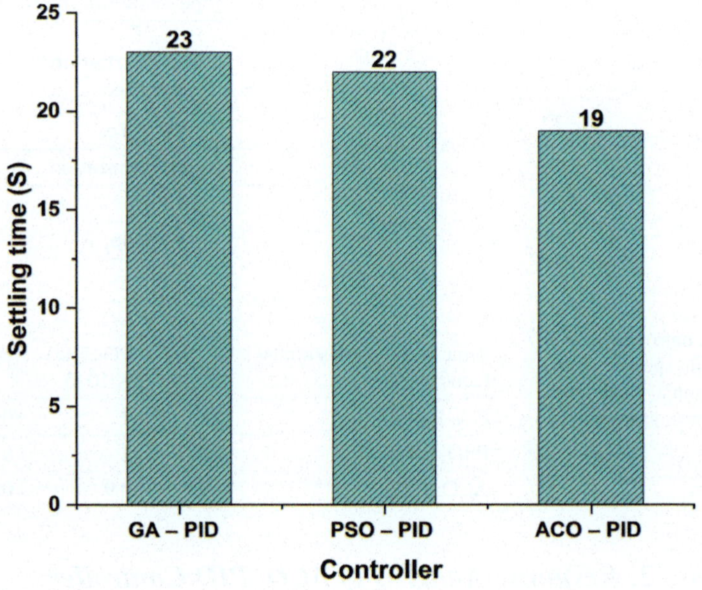

Fig. 7 Bar chart assessment of Ts (single area)

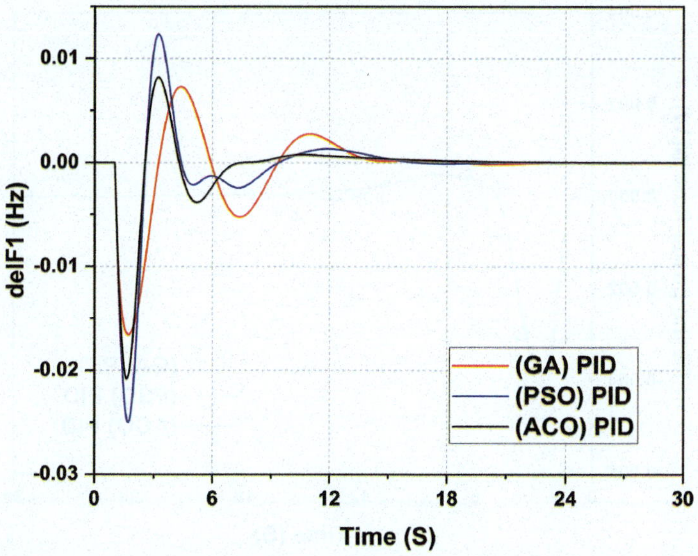

Fig. 8 Frequency oscillation of two-area thermal (area 1) comparison

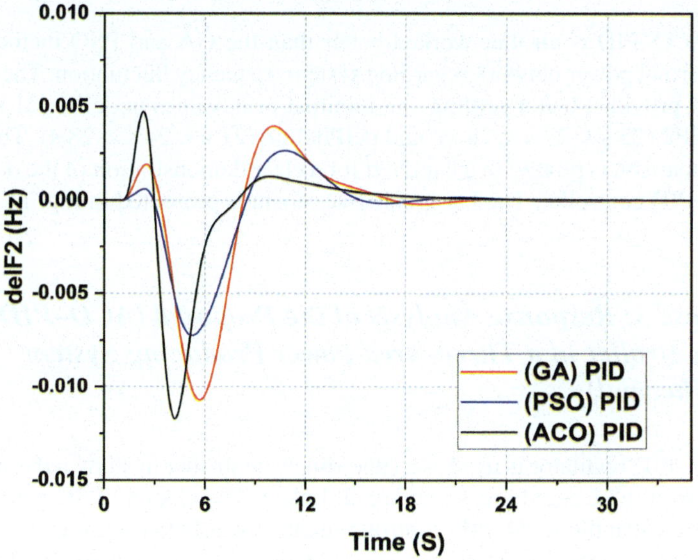

Fig. 9 Frequency oscillation of two-area thermal (area 2) comparison

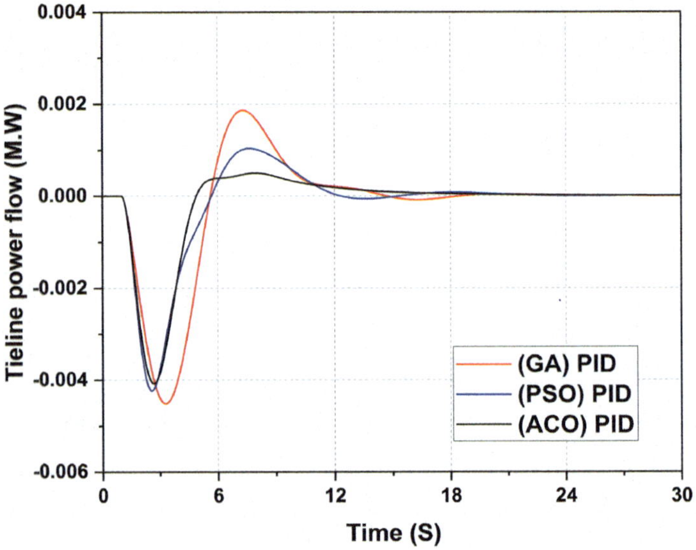

Fig. 10 Tie-line power flow deviation comparison

The ACO–PID controller worked better than the GA and PSO methods in the linked thermal power network regarding system frequency fluctuation. The proposed controller provided better settling and minimal peak values in delF1 (21 s < 23 s < 25 s), delF2 (25 s < 27 s < 28 s), and delPtieline (21 s < 24 s < 28 s). The settling time comparison is plotted in a bar chart for a clear demonstration of the domination of ACO–PID controller. The bar graph assessment is presented in Fig. 11.

4.3 Case 3: Response Analysis of the Proposed (ACO–PID) Controller in a Three-Area Power Producing System (Thermal)

The ACO–PID regulator is used as a subordinate controller to study a thermal power network with three equal areas that are all linked. The GA and PSO methods were used to show that the ACO–PID controller method was better. Figures 12, 13 and 14 show in what way the frequency deviations of areas 1, 2, and 3 compare. Figures 15, 16, and 17 also show how the power flows between two places. The nominal values from Figs. 12, 13, 14, 15, 16, 17 are given in Table 6.

The three-area interlinked thermal power producing network is studied by the developed (ACO–PID) controller for LFC. The proposed ACO–PID controller response has been compared with GA and PSO method-tuned PID controllers. The response comparison Figs. 12, 13, 14, 15, 16, 17 and Table 6 demonstrate that the

Table 5 Controlled time-domain specific parameters in the two-area thermal power system

Time domain specific parameters/ Controller	delF1			delF2			deltieline		
	$T_{settling}$ (S)	$P_{overshoot}$ (Hz) X 10^{-3}	$P_{undershoot}$ (Hz)	$T_{settling}$ (S)	$P_{overshoot}$ (Hz) X 10^{-3}	$P_{undershoot}$ (Hz)	T settling (S)	$P_{overshoot}$ (MW) X 10^{-3}	$P_{undershoot}$ (MW)
GA–PID	25	7.3	0.016	28	4	10.5	28	1.9	4.5
PSO–PID	23	12	0.025	27	2.7	7.2	24	1	4.2
ACO–PID	21	8	0.02	25	4.5	12	21	0.5	4

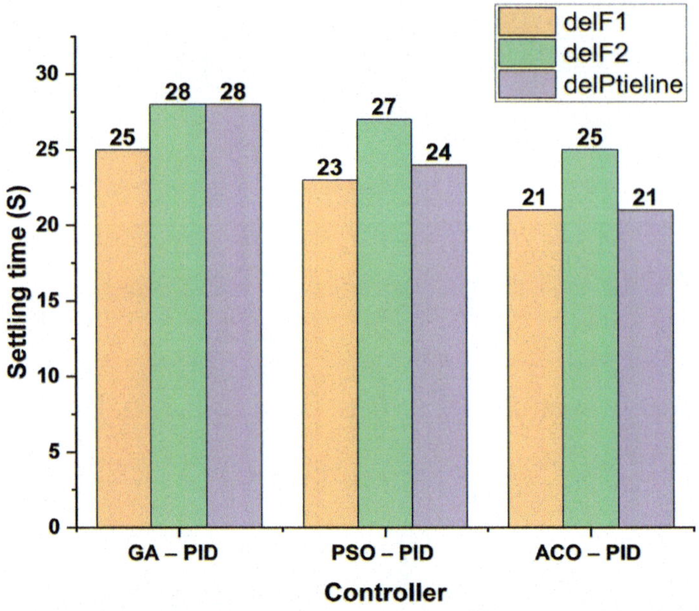

Fig. 11 Bar chart assessment of Ts (two areas)

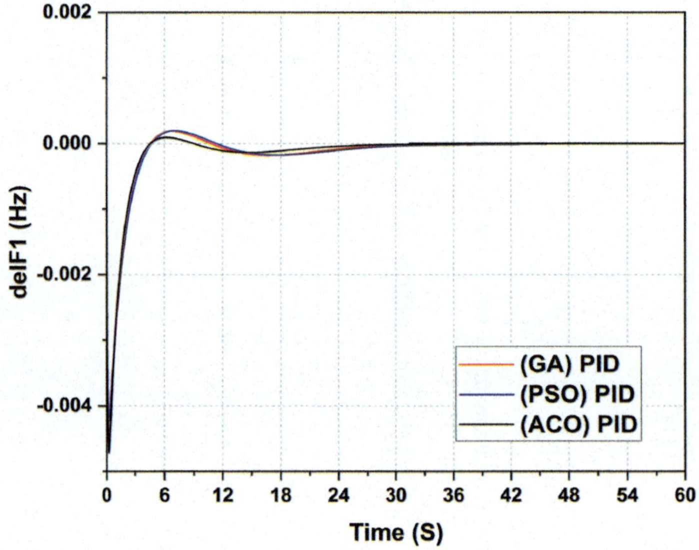

Fig. 12 Frequency oscillation of three area thermal (area 1) comparison

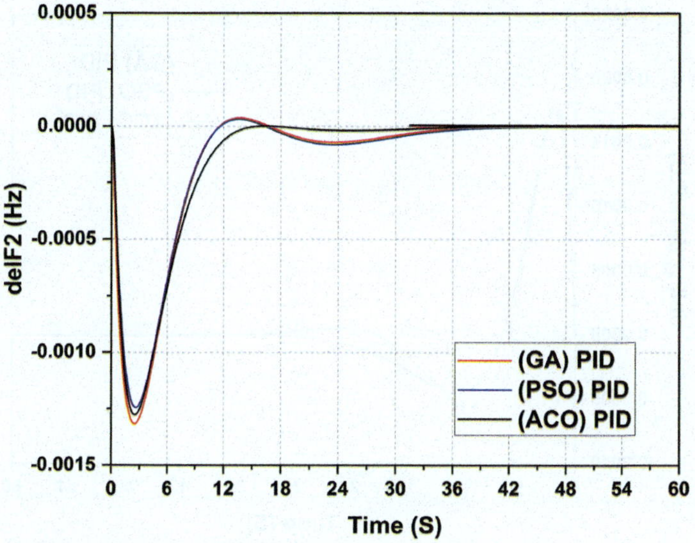

Fig. 13 Frequency oscillation of three area thermal (area 2) comparison

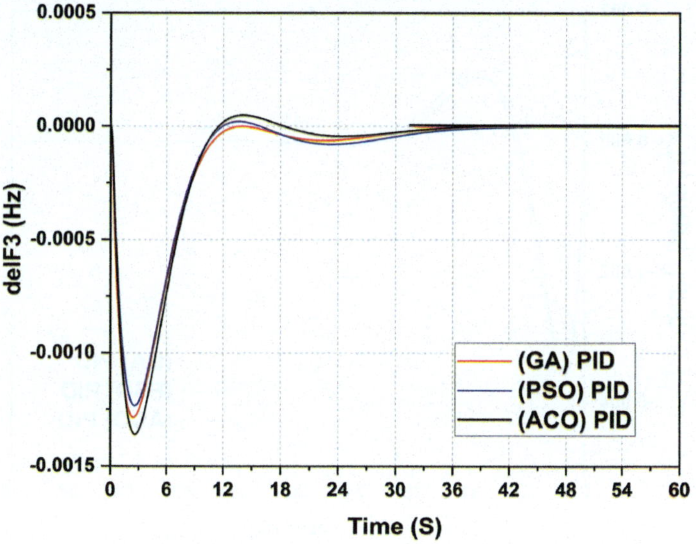

Fig. 14 Frequency oscillation of three area thermal (area 3) comparison

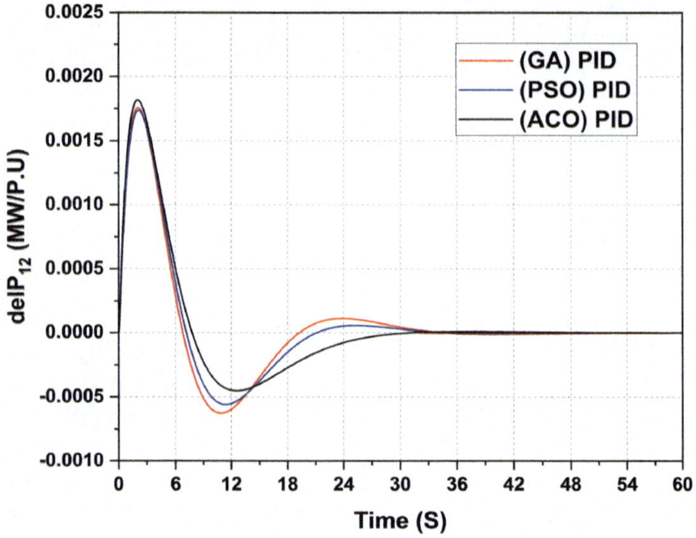

Fig. 15 Tieline power oscillation of three area thermal (areas 1 and 2) comparison

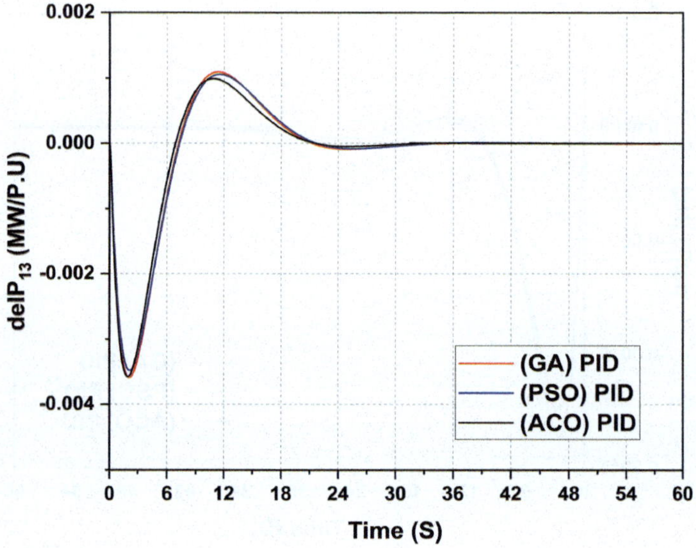

Fig. 16 Tieline power oscillation of three area thermal (areas 1 and 3) comparison

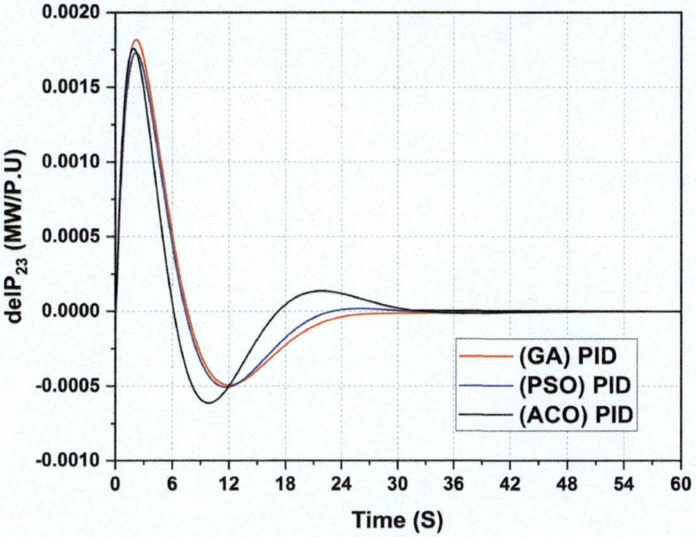

Fig. 17 Tieline power oscillation of three area thermal (areas 2 and 3) comparison

ACO technique is superior to the GA and PSO techniques. ACO controlled the oscillation of delF1 (32 s = 32 s > 28 s), delF2 (41 s > 40 s > 38 s), delF3 (38 s > 37 s > 35 s), delP12 (49 s > 46 s > 32 s), delP23 (41 s > 38 s > 35 s), and delP13 (35 s > 32 s > 29 s). Once again, the denomination of ACO–PID controller proved with a bar chart, the chart shown in Figs. 18 and 19.

Challenges and Limitations of This Study

In this chapter, the proposed system complexity has been increased for a deep analysis of the performance of the ACO technique. While increasing the size of the system, gain values are increased. The tuning of multi-parameters is a little bit tough. The controller gain parameters were tuned with the fixed boundary values.

Table 6 Controlled time-domain specific parameters in the three-area thermal power system

Timedomain specific parameters/ controller	delF1			delF2			delF3		
	$T_{settling}$ (S)	$P_{overshoot}$ (Hz) X 10^{-3}	$P_{undershoot}$ (Hz)	$T_{settling}$ (S)	$P_{overshoot}$ (Hz) X 10^{3}	$P_{undershoot}$ (Hz)	$T_{settling}$ (S)	$P_{overshoot}$ (Hz) X 10^{-3}	$P_{undershoot}$ (Hz)
GA–PID	32	0.2	4.5	41	5	13	38	1.1	3.5
PSO–PID	32	0.2	4.2	40	4	1.25	37	5	1.2
ACO–PID	28	0.1	4.5	38	0	1.25	35	5	1.3
Timedomain specific parameters/ controller	delP12			delP23			delP13		
	$T_{settling}$ (S)	$P_{overshoot}$ (M) X 10^{-3}	$P_{undershoot}$ (MW)	$T_{settling}$ (S)	$P_{overshoot}$ (M) X 10^{-3}	$P_{undershoot}$ (MW)	$T_{settling}$ (S)	$P_{overshoot}$ (M) X 10^{-3}	$P_{undershoot}$ (MW)
GA–PID	49	18	6	41	18	5	35	1.1	3.5
PSO–PID	46	17.5	5.1	38	17	5	32	1	3.5
ACO–PID	32	18	4.5	35	17	6	29	1	3.5

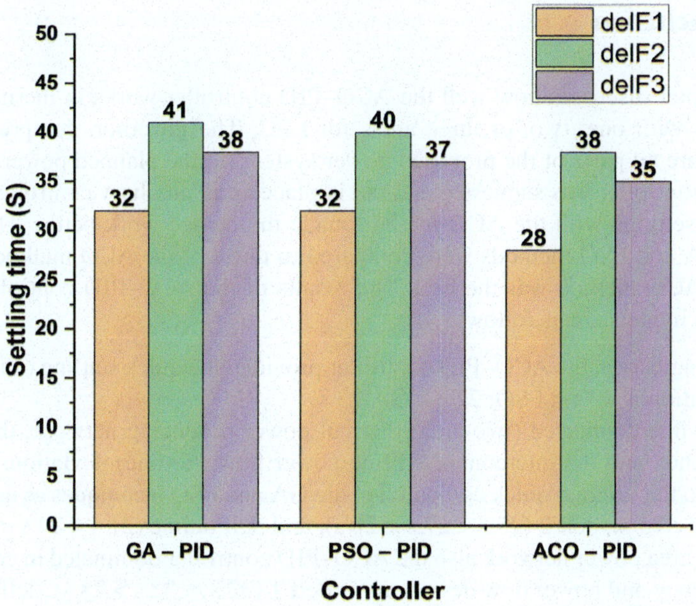

Fig. 18 Bar chart assessment of Ts (two area delF)

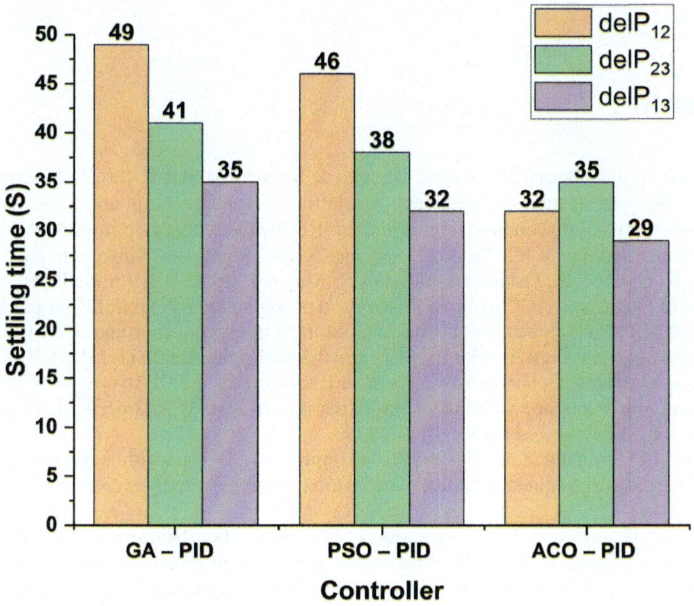

Fig. 19 Bar chart assessment of Ts (two-area delP)

5 Conclusion

This chapter discusses how well the ACO–PID controller works in thermal power networks with one, two, or three areas for LFC. The governor, compressor, and reheater are all parts of the proposed power system. In the planned power network, the PID controller was suggested and put in place as an auxiliary controller. Tuning the gain settings with the ACO method made the device work better. The results of the GA and PSO methods were compared to those of the ACO method to show that the ACO method was the best. The results of the ACO–PID controller in the proposed models are as follows:

- In a single area, the ACO–PID controller provided the quick settling of frequency oscillation is 19 s < PSO: 22 s < GA:23 s.
- In the interconnected (two-area) thermal power producing network, the system frequency and the interconnected line power flow (tie-line) deviations were set back to the original quick compared to the GA and PSO techniques as delF1 (21 s < 23 s < 25 s), delF2 (25 s < 27 s < 28 s), and delPtieline (21 s < 24 s < 28 s).
- Three area power network also the ACO–PID controller dominated in controlling frequency and power flow deviations as delF1 (32 s = 32 s > 28 s), delF2 (41 s > 40 s > 38 s), delF3 (38 s > 37 s > 35 s), delP12 (49 s > 46 s > 32 s), delP23 (41 s > 38 s > 35 s), and delP13 (35 s > 32 s > 29 s).

References

1. Boopathi D, Jagatheesan K, Anand B, Jaya J, Satheeshkumar R (2023) Investigation of a microgrid power system for frequency regulation by implementing ant colony optimization technique optimized secondary controller. In: Renewable energy optimization, planning and control: proceedings of ICRTE 2022. Springer Nature Singapore, Singapore, pp. 229–236
2. Khezri R, Oshnoei A, Oshnoei S, Bevrani H, Muyeen SM (2019) An intelligent coordinator design for GCSC and AGC in a two-area hybrid power system. Appl Soft Comput 76:491–504
3. Davtalab S, Tousi B, Nazarpour D (2022) Optimized intelligent coordinator for load frequency control in a two-area system with PV plant and thermal generator. IETE J Res 68(5):3876–3886
4. Khamies M, Magdy G, Hussein ME, Banakhr FA, Kamel S (2020) An efficient control strategy for enhancing frequency stability of multiarea power system considering high wind energy penetration. IEEE Access 8:140062–140078
5. Chen G, Li Z, Zhang Z, Li S (2019) An improved ACO algorithm optimized fuzzy PID controller for load frequency control in multi area interconnected power systems. IEEE Access 8:6429–6447
6. Murugesan D, Jagatheesan K, Boopathi D (2021) Meta-heuristic strategy planned controller for frequency supervision of integrated thermal plant with renewable source. In: 2021 IEEE 3rd PhD colloquium on ethically driven innovation and technology for society (PhD EDITS). IEEE, pp. 1–2
7. Guha D, Roy PK, Banerjee S (2018) Symbiotic organism search algorithm applied to load frequency control of multiarea power system. Energy Systems 9:439–468
8. Naidu K, Mokhlis H, Bakar AHA, Terzija V (2014) Comparative performance analysis of firefly algorithm for load frequency control in automatic generation control of interconnected reheat thermal power system

9. Kumarakrishnan V, Vijayakumar G, Jagatheesan K, Boopathi D, Anand B, Kanendra Naidu V (2022) PSO optimum design-PID controller for frequency management of single area multi-source power generating system. In: Contemporary issues in communication, cloud and big data analytics: proceedings of CCB 2020. Springer Singapore, pp. 373–383
10. Abou El Ela AA, El-Sehiemy R, Shaheen AM, Diab AEG (2020) Optimal design of PID controller based sampe-jaya algorithm for load frequency control of linear and nonlinear multiarea thermal power systems. Int J Eng Res Afr 50:79–93
11. Nguyen GN, Jagatheesan K, Ashour AS, Anand B, Dey N (2018) Ant colony optimization based load frequency control of multiarea interconnected thermal power system with governor dead-band nonlinearity. In: Smart trends in systems, security and sustainability: proceedings of WS4 2017. Springer Singapore, pp. 157–167
12. Sahu RK, Panda S, Pradhan PC (2015) Design and analysis of hybrid firefly algorithm-pattern search based fuzzy PID controller for LFC of multi area power systems. Int J Electr Power Energy Syst 69:200–212
13. Dhanasekaran B, Kaliannan J, Baskaran A, Dey N, Tavares JMR (2023) Load frequency control assessment of a PSO-PID controller for a standalone multi-source power system. Technologies 11(1):22
14. Ahmed M, Magdy G, Khamies M, Kamel S (2022) Modified TID controller for load frequency control of a two-area interconnected diverse-unit power system. Int J Electr Power Energy Syst 135:107528
15. Boopathi D, Jagatheesan K, Anand B, Samanta S, Dey N (2023) Frequency regulation of interlinked microgrid system using mayfly algorithm-based PID controller. Sustainability 15(11):8829
16. Fathy A, Kassem AM, Abdelaziz AY (2020) Optimal design of fuzzy PID controller for deregulated LFC of multiarea power system via mine blast algorithm. Neural Comput Appl 32:4531–4551
17. Elsisi M, Aboelela M, Soliman M, Mansour W (2018) Design of optimal model predictive controller for LFC of nonlinear multiarea power system with energy storage devices. Electric Power Components Syst. 46(11–12):1300–1311
18. Khamies M, Magdy G, Selim A, Kamel S (2022) An improved Rao algorithm for frequency stability enhancement of nonlinear power system interconnected by AC/DC links with high renewables penetration. In: Neural computing and applications, pp 1–29
19. Boopathi D, Jagatheesan K, Anand B, Kumarakrishnan V. Samanta S (2022) Effect of sustainable energy sources for load frequency control (LFC) of single-area wind power systems. In: Industrial transformation. CRC Press, pp 87–98
20. Khamies M, Magdy G, Kamel S, Elsayed SK (2021) Slime mold algorithm for frequency controller design of a two-area thermal-PV power system. In: 2021 IEEE international conference on automation/XXIV congress of the chilean association of automatic control (ICA-ACCA). IEEE, pp 1–7
21. Omar M, Ghany AMA, Bendary F (2015) Harmony search based PID for multi area load frequency control including boiler dynamics and nonlinearities. WSEAS Trans Circuits Syst 14:407–414
22. Ali T, Malik SA, Daraz A, Aslam S, Alkhalifah T (2022) Dandelion optimizer-based combined automatic voltage regulation and load frequency control in a multiarea, multisource interconnected power system with nonlinearities. Energies 15(22):8499
23. Boopathi D, Jagatheesan K, Sourav Samanta AB, Satheeshkumar R (2023) Energy storage units for frequency management in nuclear generators-based power system. In: Energy storage technologies in grid modernization, p. 105
24. Bhagat SK, Behera S, Saha A, Bal T, Chiranjeevi T, Devarapalli R, Babu NR (2023) Bird swarm algorithm optimized TIDD controller for multiarea load frequency control application. SN Comput Sci 4(2):138
25. Kumarakrishnan V, Vijayakumar G, Boopathi D, Jagatheesan K, Saravanan S, Anand B (2022) Frequency regulation of interconnected power generating system using ant colony optimization technique tuned PID controller. In: Control and measurement applications for smart grid: select proceedings of SGESC 2021. Springer Nature Singapore, Singapore, pp 129–141

26. Dhanasekaran B, Siddhan S, Kaliannan J (2020) Ant colony optimization technique tuned controller for frequency regulation of single area nuclear power generating system. Microprocess Microsyst 73:102953
27. Jagatheesan K, Samanta S, Boopathi D, Anand B (2021) Frequency stability analysis of microgrid interconnected thermal power generating system with GWO tuned PID controller. In: 2021 9th IEEE international conference on power systems (ICPS). IEEE, pp 1–5
28. Boopathi D, Jagatheesan K, Samanta S, Anand B, Jaya J (2023) Application of genetic algorithm-based controllers in wind energy systems for smart energy management. In: Applied genetic algorithm and its variants: case studies and new developments. Springer Nature Singapore, Singapore, pp 139–160
29. Jagatheesan K, Anand B, Dey N, Ashour AS (2018) Effect of SMES unit in AGC of an interconnected multi-area thermal power system with ACO-tuned PID controller. In: Advancements in applied metaheuristic computing. IGI Global, pp 164–184
30. Moraes CA, De Oliveira EJ, Khosravy M, Oliveira LW, Honório LM, Pinto MF (2020) A hybrid bat-inspired algorithm for power transmission expansion planning on a practical Brazilian network. In: Applied nature-inspired computing: algorithms and case studies, pp 71–95
31. Reddy S, Panwar L, Panigrahi BK, Kumar R, AlSumaiti AS (2020) An application of binary grey wolf optimizer (BGWO) variants for unit commitment problem. In: Applied nature-inspired computing: algorithms and case studies, pp. 97–127

Chapter 11
Frequency-Based Optimization of Truss Dome Structures Using Ant Colony Optimization ($ACO_{\mathbb{R}}$) with Multi-trail Pheromone Memory

Salar Farahmand-Tabar

1 Introduction

All engineers share the goal of achieving their objectives with the minimum computing effort and cost possible. These objectives include providing humans with a variety of housing options, electrical appliances, and mechanical tools. Two categories of optimization techniques can now accomplish these objectives: local optimizers, which employ mathematical programming methods and utilize gradient information to explore the area around a starting point [1], and global optimizers known as metaheuristics, which are suitable for global searches and do not require continuous cost functions and variables [2–14].

Truss structural analysis is a computationally expensive task in real-world projects. Two critical needs form the basis for truss design: determining the optimal geometrical arrangement for bars and nodes and calculating the most suitable cross-sections. Layout optimization, which involves simultaneously optimizing size (e.g., cross-sectional areas) and shape (e.g., node coordinates), presents a typical multi-modal and highly nonlinear challenge. Due to the differences between the two kinds of variables, the layout optimization of a truss is particularly demanding. The geometry of the structure is generally dictated by engineering requirements and influenced by aesthetic, economic, construction-related, and environmental factors. Additionally, the size of the elements is determined based on functional and failure criteria.

Trusses are skeletal structures that are widely optimized while considering frequency constraints. Various optimization techniques have been applied to the size optimization of trusses, such as the modified simulated annealing algorithm [15],

S. Farahmand-Tabar (✉)
Department of Civil Engineering, Faculty of Engineering, University of Zanjan, Zanjan, Iran
e-mail: farahmandsalar@znu.ac.ir; farahmandsalar@gmail.com

N. Dey (ed.), *Applications of Ant Colony Optimization and its Variants*, Springer Tracts in Nature-Inspired Computing, https://doi.org/10.1007/978-981-99-7227-2_11

singular solutions [16], teaching learning-based optimization [17], improved water swirl algorithm [18], and a comparative study of improved DE with other algorithms [19]. Some authors have considered multiple frequency constraints in their examples [20, 21]. Hybrid and enhanced algorithms have also been proposed to address the frequency-based size optimization of trusses, including the hybrid CSS-MBLS algorithm [22], hybrid Jaya algorithm [23], hybrid DE and Jaya algorithm [24], and hybrid OC-GA approach [25].

In recent studies, various advancements have been employed in frequency-based truss optimization. These advancements include enhancements in differential evolution using nearest neighbor comparison and adaptive directional mutation [26], improvements in the DE algorithm through an adaptive mutation scheme [27], the application of a fruit fly algorithm with the strategy of adaptive vision search [28], utilization of multiclass teaching–learning-based optimization [29], enhancements in the symbiotic organism search [30], a more effective particle swarm optimization method incorporating gradient directions [31], and modifications in the subpopulation teaching–learning-based optimization [32].

However, many fundamental optimization algorithms encounter limitations, such as slow convergence and a tendency to get stuck in local optima. To overcome these challenges, researchers have proposed various mechanisms to enhance algorithm performance [33, 34]. In this chapter, the concept of multi-trail pheromone memory is introduced for ant colony optimization. This concept involves a dedicated memory component that stores and exchanges the best solutions found during the process of optimization. To evaluate the efficiency of using memory in this algorithm, experiments are conducted using various benchmark truss dome structures, and their effectiveness is compared against standard versions of the algorithm.

2 Related Works on Memory-Enhanced Metaheuristics

Memory-enhanced metaheuristics have gained increasing attention in recent years due to their ability to enhance the performance of optimization algorithms. These methods are designed to incorporate memory mechanisms into conventional metaheuristics, such as differential evolution, genetic algorithms, ant colony optimization, particle swarm optimization, artificial bee colony, gray wolf optimizer, and whale optimization algorithm. Incorporating memory mechanisms into the search process can significantly improve its effectiveness. By storing and utilizing information about previously visited solutions, these mechanisms enable the algorithm to explore the search space more efficiently, leading to faster convergence toward better solutions. In this section, a comprehensive overview of the background studies on memory-enhanced metaheuristics is provided, covering a range of optimization methods and their applications in various fields. Table 1 summarizes the methods and their applications discussed in this section.

The applications range from global and engineering optimization to machine learning-related applications. The incorporation of memory mechanisms into these

Table 1 Background studies on memory-assisted metaheuristics

Method	Application	Refs.	Method	Application	Refs.
GA	Laminate composites	[35]	SAGA	Multimodal optimization	[36]
NSGA II	Dynamic Problems	[37]	EA	Job Shop Scheduling	[38]
HDE	Continuous problems	[39]	PSO	Dynamic optimization	[40]
QPSO	Large scale problems	[41]	ACO	Dynamic vehicle routing	[42]
DA	Engineering problems	[43]	IFFOA	Time series forecasting	[44]
GWO	Global optimization	[45]	AFSA	Multiextreme value func	[46]
SCA	Global optimization	[47]	CFA	Global optimization	[48]
WOA	Smooth path planning	[49]	KMTOA	Global optimization	[50]
MMGSO	Benchmark functions	[51]	OWMA	Global optimization	[52]
BBBC	Data clustering	[53]	GD	Global optimization	[54]
HMBOA	Dynamic environments	[55]	P-PDA	Image processing	[56]
CSA	Global optimization	[57]	TRA	Unconstrained opt	[58]
CA	Combinatorial opt	[59]	AIS	Dynamic optimization	[60]

metaheuristics has been found to improve the algorithms' performance, allowing for more effective exploration of the search space and faster convergence to better solutions. Overall, the use of memory-enhanced metaheuristics is a promising area of research with the potential to enhance optimization performance across a wide range of applications.

3 Proposed Ant Colony Optimization Algorithms

3.1 Standard ACO for Continuous Domain ($ACO_{\mathbb{R}}$)

Ant colony optimization for a continuous domain ($ACO_{\mathbb{R}}$), introduced by Socha et al. [61], is an extension of the ant colony optimization algorithm designed specifically for real-variable optimization. Similar to the traditional ACO algorithm, $ACO_{\mathbb{R}}$ updates its pheromone information according to the components of promising solutions discovered during the search process. However, there is a notable distinction between these two approaches. In the standard ACO, the solutions discovered by the ants are discarded after updating the pheromone information. In contrast, $ACO_{\mathbb{R}}$ follows a different strategy by maintaining a solution archive, denoted as T, which keeps track of a predetermined number of candidate solutions.

Essentially, the solution archive T used by $ACO_{\mathbb{R}}$ is represented as a table that stores R previous candidate solutions. This archive plays a crucial role in the algorithm because it allows $ACO_{\mathbb{R}}$ to retain and evaluate these candidate solutions,

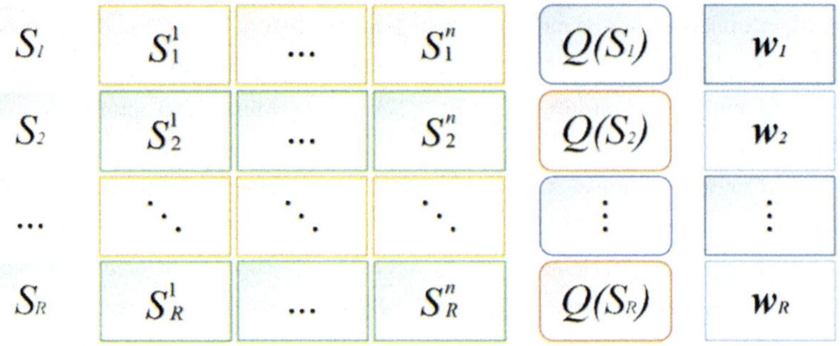

Fig. 1 Solution archive T saved by $ACO_{\mathbb{R}}$

thereby contributing to the overall effectiveness of the optimization process. For a clearer understanding of the structure of the solution archive T, please refer to Fig. 1.

In archive T, the elements denoted as S_a represent potential solutions for the problem. Each solution is a real-valued vector in n-dimensional space, where n is the number of variables in the problem. Precisely, $S_a = (S_a^1, S_a^2, \ldots, S_a^n)$, and $S_{a,j}$ represents the value of the j th variable in the a th solution within the archive table T. The solutions in T are arranged in descending order of their quality, denoted as $Q(S_1) \geq Q(S_2) \geq \cdots \geq Q(S_R)$. Additionally, each solution S_a in archive T is associated with a weight W_a, which relates to its quality, resulting in $w_1 \geq w_2 \geq \cdots \geq w_R$.

The process of $ACO_{\mathbb{R}}$ includes two phases: construction of the solution and updating the pheromone. The core aspect of $ACO_{\mathbb{R}}$ lies in the phase of constructing a solution, where each ant i creates a candidate solution S_i. During the construction of its solution S_i, ant i draws influence from one of the R solutions existing in the archive table T. At the beginning, the ant chooses one of the R solutions from the archive table T probabilistically based on specific criteria.

$$p(S_a) = \frac{w_a}{\sum_{r=1}^{R} w_r} \tag{1}$$

In Eq. (1), it is evident that the likelihood of choosing a solution S_a is directly linked to its weight w_a. Crucially, the archive T is organized based on the quality of solutions, resulting in each solution S_a being assigned a rank within the range of $[1, R]$. The weights w_a utilized in Eq. (1) are generated during each iteration through the following outlined process:

$$w_a = \frac{1}{\sigma\sqrt{2\pi}} e^{\frac{(a-\mu)^2}{2\sigma^2}} \tag{2}$$

It is evident from Eq. (2) that the weight w_a is determined by evaluating the Gaussian function with the argument a, in which the mean (μ) is fixed at 1, and the standard deviation (σ) is computed using the following formula:

$$\sigma = qR \tag{3}$$

The algorithm incorporates an additional external control parameter denoted as q. By leveraging the characteristics of the Gaussian function, it becomes evident that smaller values of q result in better-ranked solutions, with higher weights ω. Consequently, the algorithm tends to adopt a more exploitative approach.

During a particular iteration, let S_a represent the solution chosen from archive T by ant i, following Eq. (1). Subsequently, ant i generates the j th element of its solution S_i^j by sampling from the Gaussian probability density function (PDF). The fundamental principle at the core of $ACO_{\mathbb{R}}$ involves a shift from utilizing a discrete probability distribution to employing a continuous one, specifically, the Gaussian PDF.

$$S_i^j \sim N\left(\mu_a^j, \sigma_a^j\right) \tag{4}$$

In this context, $N(\mu, \sigma)$ represents the Gaussian PDF with the mean μ_a^j being equal to the value of the j th element in the chosen solution, S_a^j, and the standard deviation, σ_a^j.

$$\sigma_a^j = \xi \sum_{r=1}^{R} \frac{\left|S_a^j - S_r^j\right|}{R - 1} \tag{5}$$

The parameter ξ serves as an external factor in the algorithm, playing a role similar to evaporation in the standard ACO algorithm. Its value influences the degree to which the search is biased toward the vicinity of the solutions stored in the archive within the search space. When ξ is set to lower values, there is a stronger bias toward the solutions in the archive, which leads to faster convergence of the algorithm.

Once the solution construction phase is finished, the pheromone information (archive T) undergoes an update. This entire process is iterated until the desired stopping criteria are satisfied. The flowchart of $ACO_{\mathbb{R}}$ is depicted in Fig. 2.

3.2 *Enhancement Strategy*

3.2.1 Multi-trail Pheromone Memory

To improve algorithm performance without introducing additional computational costs, the multi-elite strategy can be employed. This approach involves using a separate memory to store historical best solutions and their corresponding fitness values.

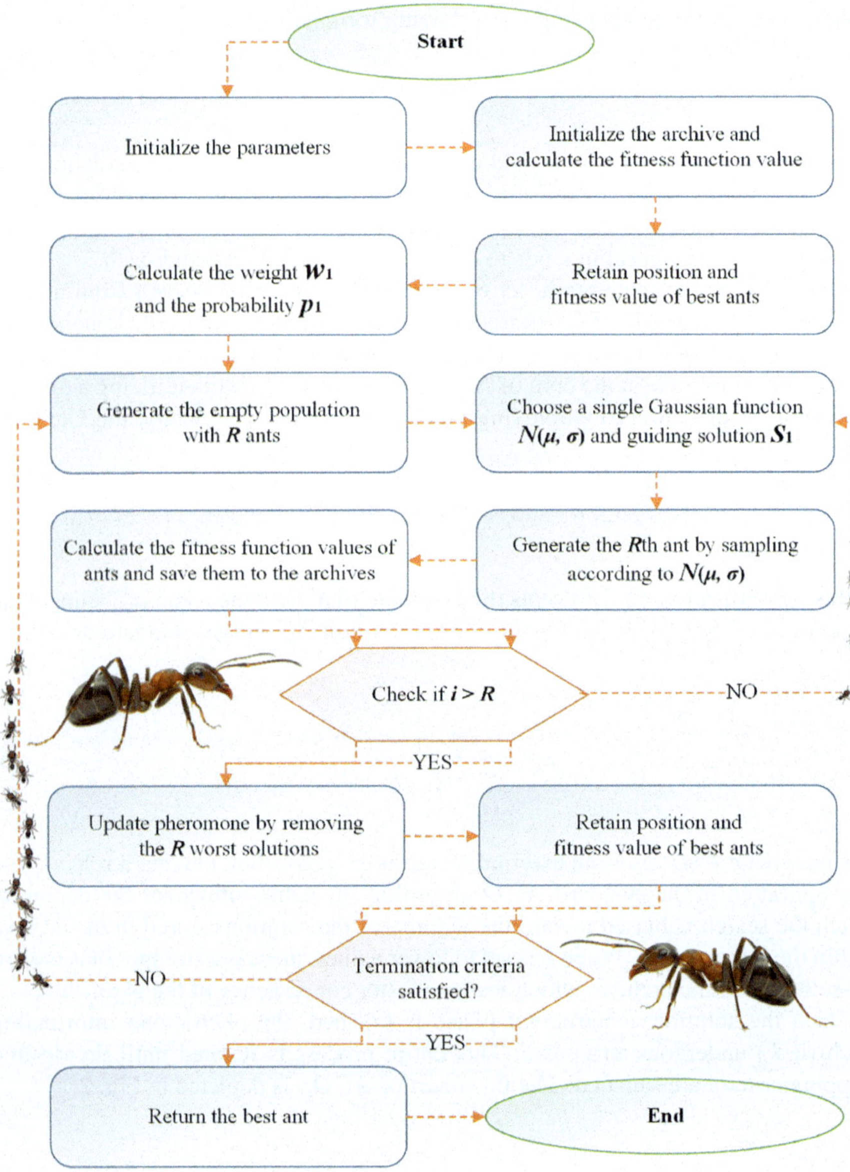

Fig. 2 Flowchart of $ACO_{\mathbb{R}}$

Unlike the elitist strategies commonly utilized by standard algorithms such as genetic algorithms, which rely on a single elite solution, the multi-elite strategy preserves a specific number of the best solutions found thus far based on the designated memory size. During each iteration, these elite solutions are replaced with an equal number of the worst solutions. In this chapter, the memory size, which represents the number of elite solutions stored, is set as 20% of the total population number. This multi-elite strategy allows for efficient retention of the most promising solutions, contributing to enhanced algorithm performance while avoiding excessive computational overhead. By incorporating this memory into each iteration (Algorithm 1), undesirable and poor-performing solutions can be replaced with the desired elite solutions. The memory is continuously updated with new solutions, and if the fitness of the new solutions surpasses that of the stored elite solutions, they are exchanged, along with their corresponding fitness values. This modification enables the algorithm to avoid suboptimal solutions and achieve faster and more efficient convergence.

Algorithm 1. Pseudocode of Assigning Memory

```
Start
% Saving more than one global best and related solutions
for i = 1:NU
  for j = 1:NU/5                % Memory size:20 percent of search agents
     if Fit(i) < Fit_M(j)
        U_M(j,:) = U(i,:);      % Best solutions of the memory
        Fit_M(j) = Fit(i);      % Best fitness values of the memory
     end
  end
end
% Worst 20% of search agents are changed in the main loop of the algorithm
```

4 State of the Optimization Problem

This section focuses on the optimization of truss structures as benchmark examples. The study includes two types of trusses: 120-bar and 52-bar truss domes. In these examples, the problem's objective function is to optimize the structural weight of the truss, making weight reduction the primary goal of the optimization process. Nodal displacements and element stresses serve as the optimization constraints. The problem may be described in the following way:

$$Minimize: W = \sum_{i=1}^{n} \gamma_i l_i A_i, \quad i = 1, 2, ..., n \tag{6}$$

$$subject\ to: g_o^f = \frac{f_o}{\overline{fo}}, \quad o = 1, 2, ..., n \tag{7}$$

The given equation represents the structural design optimization problem, where various parameters are involved:

A_i, l_i, and γ_i are the cross-sectional area, length, and material density of the i th element, respectively; n and m denote the total number of elements and nodes in the truss structure, respectively; g_o^f corresponds to the frequency constraints of the structure; and $\overline{f_O}$ represents acceptable values for stress, displacement, frequency, and buckling constraints.

The penalty function considered for this optimization problem is as follows:

$$PW_m = \sum_{f=1}^{N} W \times \left(1 + C\left(\sum_{Iel=1}^{Nel} G_{Iel}^{i,freq.}\right)\right)^T \tag{8}$$

In this context, the optimization involves two objective functions: PW_m and W_m. The former accounts for the objective function with the penalty term incorporated, while the latter represents the objective function solely based on the structural weight. The penalty function's coefficient and exponent are denoted as C and T, respectively. These parameters play a pivotal role in the penalty function, shaping its impact on the optimization process and influencing how constraint violations are handled.

5 Results and Discussion on the Case Studies

5.1 The 120-Bar Truss Dome

A 120-bar truss dome [62] is an example of size optimization, and it is shown in Fig. 3. Nodes 1, 2, and 13 are each attached with nonstructural weights of 500 kg, 3,000 kg, and 100 kg, respectively. Table 2 contains the information for the problem, including the properties of the material, limitations of the design variable, and constraints for frequencies.

The statistical results of the considered techniques are presented in Table 3. The results of $ACO_{\mathbb{R}} - Mem.$ are compared with standard $ACO_{\mathbb{R}}$ and a method from the literature [62] called orthogonal multi-gravitational search ($OMGSA$). According to the table, the optimum weight achieved by $ACO_{\mathbb{R}} - Mem.$ with a value of 8708.8564 kg is lower than those of $ACO_{\mathbb{R}}$ and $OMGSA$ with values of 8716.4805 kg and 8,724.97 kg, respectively.

The convergence history illustrated in Fig. 4 demonstrates that $ACO_{\mathbb{R}} - Mem.$ has better local and global search capabilities than $ACO_{\mathbb{R}}$. The initial frequencies of the 120-bar truss dome at the achieved optimum design are shown in Table 3. This shows that none of the frequency restrictions are violated (Fig. 5).

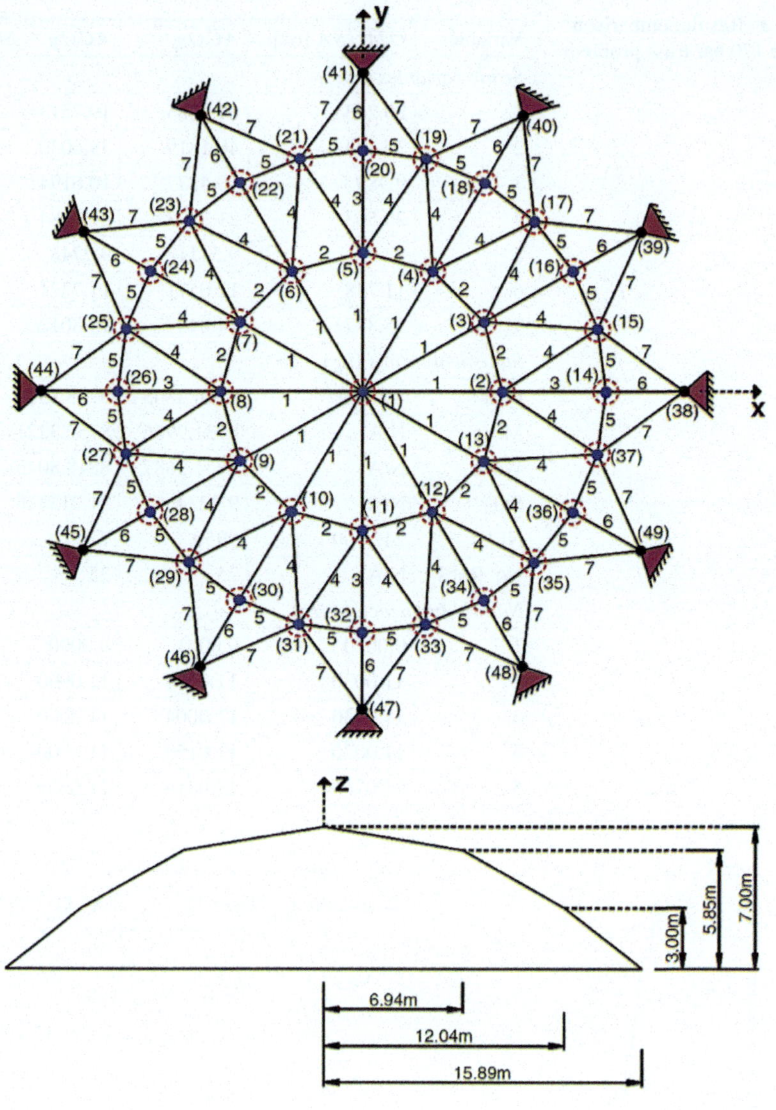

Fig. 3 The 120-bar truss dome

Table 2 Data for design of 120-bar planar truss

Parameters	Value
Elasticity modulus $E(N/m^2)$	2.1×10^{11}
Density of material $\rho(kg/m^3)$	7971.810
Additional Mass (kg)	$m_1 = 3000$, $m_2 = 500$, $m_3 = 100$,
Permitted range of cross-sections (m^2)	$0.0001 \leq A \leq 0.01293$
Constraints on initial frequencies (Hz)	$\omega_1 \geq 9, \omega_2 \geq 11$

Table 3 Result comparison for the 120 bar truss problem

Variables	$OMGSA$ [62]	$ACO_{\mathbb{R}}$	$ACO_{\mathbb{R}} - Mem.$
Sizing variables (cm^2)			
1	20.263	19.6885	19.7513
2	39.294	40.1819	39.6030
3	9.989	10.4877	10.6194
4	20.563	21.2283	21.2884
5	9.603	9.54712	9.7743
6	11.738	12.0571	11.7321
7	15.877	14.8447	14.8012
Statistical results (kg)			
Best	8,724.97	8716.4805	8708.8564
Mean	N/A	8731.9966	8752.3221
Worst	N/A	8751.2687	8818.8045
Std	N/A	9.2234	31.913318
NFEs	242,700	9950	7700
No. Runs	N/A	25	25
Natural frequencies (Hz)			
1	9.0020	9.0001	9.0000
2	11.0030	11.0004	11.0000
3	11.0030	11.0004	11.0000
4	11.0070	11.0159	11.0000
5	11.0760	11.0816	11.0666

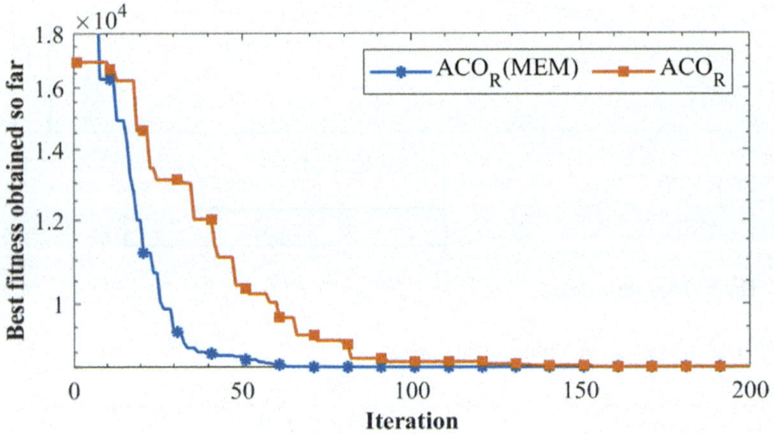

Fig. 4 Convergence history of 120-bar truss

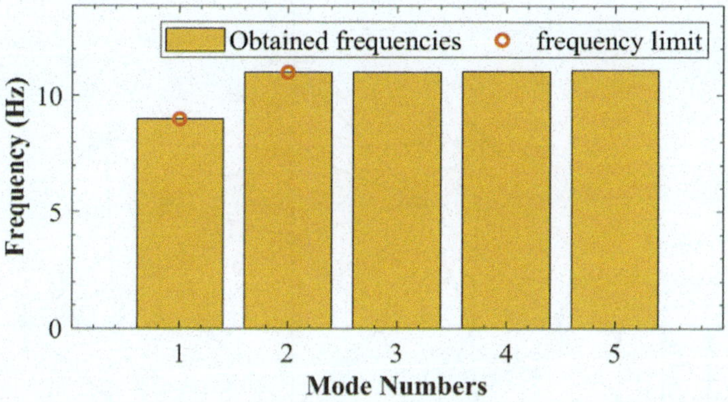

Fig. 5 The achieved frequencies and their limits in 120-bar dome

5.2 *The 52-Bar Truss Dome*

The 52-bar dome truss [63] shown in Fig. 6 is another example within the category of size and shape optimization. According to Table 4, all bars are divided into eight groups. Each free node is permitted to displace up to 2 m in both positive and negative directions from its original location while adhering to the constraint of maintaining structural symmetry. Consequently, there are eight size variables and five shape variables. Table 5 contains the problem's data, including the properties of the material, limitations of the design variables, and constraints of the frequencies.

Table 6 lists the results of the problem from the current study and earlier investigations. The findings show that the $ACO_{\mathbb{R}} - Mem.$ with the optimum weight of 193.8140 kg is superior in comparison to the $ACO_{\mathbb{R}}$ and another method from the literature with values of 201.6796 kg and 194.85 kg, respectively.

Figure 7 depicts the convergence history of the best results obtained using the proposed method. Table 6 also provides information on the initial natural frequencies of the 52-bar dome truss structure, which were determined through the optimal design. The initial two frequencies of the problem continue to meet the constraints, and considering the optimum values (193.8140 kg), it can be demonstrated that $ACO_{\mathbb{R}} - Mem.$ Converges better and faster than the other methods. The achieved initial frequencies are illustrated in Fig. 8 considering the permitted values. Additionally, Fig. 9 depicts the final configuration of the 52-bar dome truss structure resulting from the optimization process.

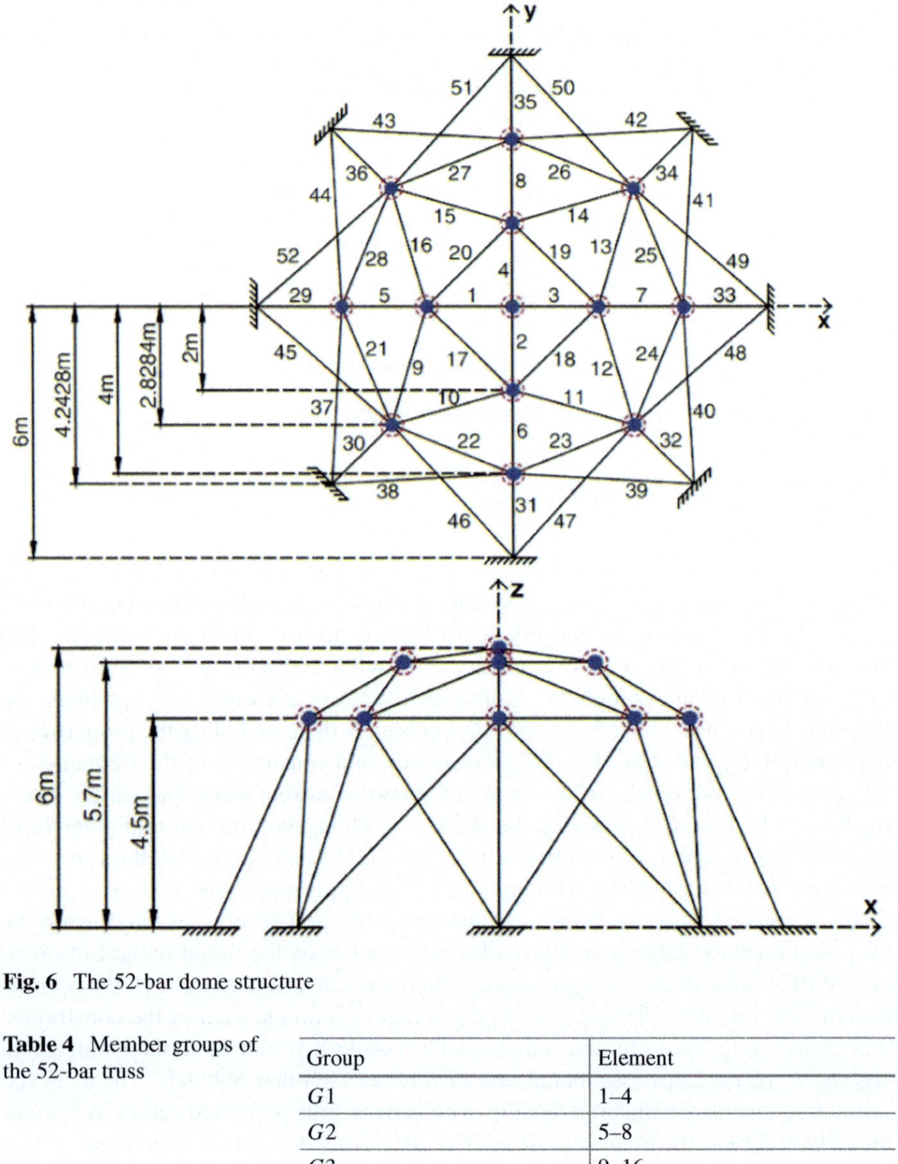

Fig. 6 The 52-bar dome structure

Table 4 Member groups of the 52-bar truss

Group	Element
*G*1	1–4
*G*2	5–8
*G*3	9–16
*G*4	17–20
*G*5	21–38
*G*6	29–36
*G*7	37–44
*G*8	45–52

Table 5 Data for design of 52-bar dome truss

Parameters	Value
Elasticity modulus $E(N/m^2)$	2.1×10^{11}
Density of material $\rho(kg/m^3)$	7800
Additional Mass (kg)	50
Permitted range of cross-sections (m^2)	$0.0001 \le A \le 0.001$
Constraints on initial frequencies (Hz)	$\omega_1 \le 15.961, \omega_2 \ge 28.648$

Table 6 Result comparison for the 52-bar dome problem

Variables	FA [63]	$ACO_{\mathbb{R}}$	$ACO_{\mathbb{R}} - Mem.$
Layout variables (m)			
Z_A	5.9362	4.0135	5.7431
X_B	2.2416	2.2666	2.1830
Z_B	3.7309	3.7074	3.7484
X_F	3.963	3.9167	3.9190
Z_F	2.500	2.5031	2.5314
Sizing variables (cm^2)			
A_1	1.0001	1.0000	1.0000
A_2	1.1654	1.0117	1.3154
A_3	1.2323	1.3318	1.2641
A_4	1.4323	1.6786	1.2694
A_5	1.3901	1.2168	1.3979
A_6	1.0001	1.0083	1.0021
A_7	1.6024	1.6559	1.4982
A_8	1.4131	1.6822	1.5091
Statistical results (kg)			
Best	194.85	201.6796	193.8140
Mean	N/A	207.6520	226.0333
Worst	196.85	238.8146	424.2742
Std	2.38	7.4716	53.1658
NFEs	7,500	9950	9700
No. Runs	N/A	25	25
Natural frequencies (Hz)			
1	11.4339	15.2475	12.1303
2	28.6480	28.6751	28.6480
3	28.6480	28.6751	28.6480
4	28.6482	29.1106	28.6481
5	28.6848	30.0892	29.4305

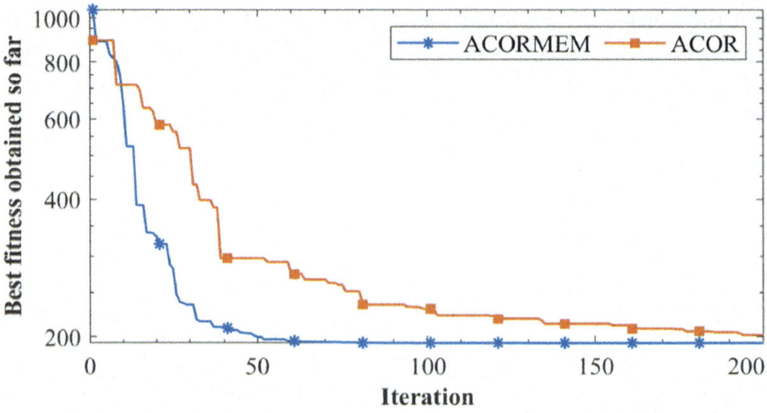

Fig. 7 Convergence history of 52-bar truss

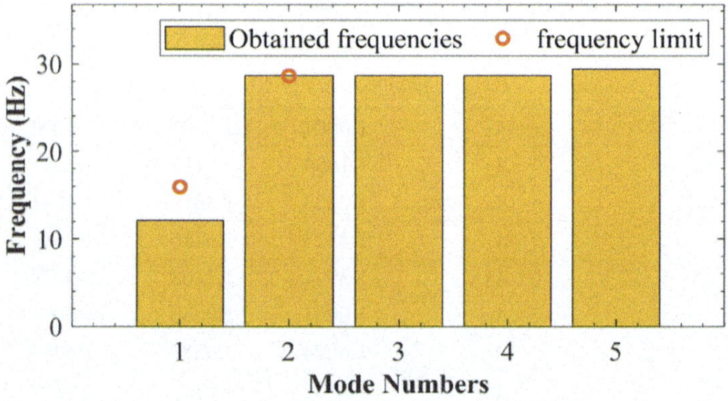

Fig. 8 The achieved frequencies and their limits in 52-bar dome

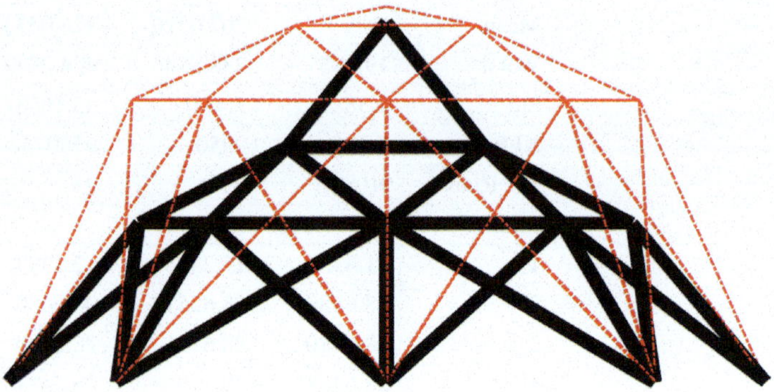

Fig. 9 The optimum shape of 52-bar dome using $ACO_{\mathbb{R}} - Mem$

6 Conclusions

In this chapter, the effectiveness of utilizing memory-based techniques in enhancing the efficiency of the ant colony optimization (ACO) algorithm has been investigated. By adding a memory component to the basic optimization algorithm, several best solutions from previous iterations are saved to be exchanged with several worst solutions in the next iterations. This strategy in ACO is called a strategy of multi-trail pheromone memory. The proposed memory implementation allowed the algorithm to avoid the worst solutions by exchanging them (20% of the population size) with the global best ones from the memory. In this chapter, the memory-enhanced ACO for the continuous domain has been implemented to optimize benchmark engineering examples, such as the size and shape optimization of truss domes. The results show that the algorithm with multi-trail pheromone memory outperforms the standard method in terms of convergence and solution reliability. The case study examples of truss domes demonstrated that the proposed method can effectively optimize problems with various variables. Therefore, utilizing the proposed technique can be a promising approach to enhance the efficiency of the algorithms to solve complex engineering problems.

References

1. Farahmand-Tabar S, Abdollahi F, Fatemi M (2023) Robust conjugate gradient methods for non-smooth convex optimization and image processing problems. In: Kulkarni AJ, Gandomi AH (eds) Handbook of formal optimization. Springer, Singapore. https://doi.org/10.1007/978-981-19-8851-6_42-1
2. Farahmand-Tabar S, Ashtari P, Babaei M (2023) Dynamic intelligence of self-organized map in the frequency-based optimum design of structures. In: Kulkarni AJ, Gandomi AH (eds) Handbook of formal optimization. Springer, Singapore. https://doi.org/10.1007/978-981-19-8851-6_45-1
3. Farahmand-Tabar S, Shirgir S (2023) Boosting the efficiency of metaheu-ristics through opposition-based learning in optimum locating of control systems in tall buildings. In: Kulkarni AJ, Gandomi AH (eds) Handbook of formal optimization. Springer, Singapore. https://doi.org/10.1007/978-981-19-8851-6_37-1
4. Farahmand-Tabar S (2023) Memory-driven metaheuristics: improving op-timization performance. In: Kulkarni AJ, Gandomi AH (eds) Handbook of formal optimization. Springer, Singapore. https://doi.org/10.1007/978-981-19-8851-6_38-1
5. Farahmand-Tabar S, Rashid TA (2023) Steel plate fault detection using the fitness dependent optimizer and neural networks. In: Kulkarni AJ, Gandomi AH (eds) Handbook of formal optimization. Springer, Singapore. https://doi.org/10.1007/978-981-19-8851-6_41-1
6. Farahmand-Tabar S, Shirgir S (2023) Synergistic collaboration of motion-based metaheuristics for the strength prediction of cement-based mortar materials using TSK model. In: Kulkarni AJ, Gandomi AH (eds) Hand-book of formal optimization. Springer, Singapore. https://doi.org/10.1007/978-981-19-8851-6_43-1
7. Farahmand-Tabar S, Shirgir S (2023) Positron-enabled atomic orbital search algorithm for improved reliability-based design optimization. In: Kulkarni AJ, Gandomi AH (eds) Handbook of formal optimization. Springer, Singapore. https://doi.org/10.1007/978-981-19-8851-6_44-1

8. Farahmand-Tabar S, Sadrekarimi N (2023) Overcoming constraints: the critical role of penalty functions as constraint handling methods in structural optimization. In: Kulkarni AJ, Gandomi AH (eds) Handbook of formal optimization. Springer, Singapore. https://doi.org/10.1007/978-981-19-8851-6_40-1
9. Farahmand-Tabar S, Babaei M (2023) Memory-assisted adaptive multiverse optimizer and its application in structural shape and size optimization. Soft Comput. https://doi.org/10.1007/s00500-023-08349-9
10. Farahmand-Tabar S (2023) Genetic algorithm and accelerating fuzzifica-tion for optimum sizing and topology design of real-size tall building systems. In: Dey N (eds) Applied genetic algorithm and its variants. Springer Tracts in nature-inspired computing. Springer, Singapore. https://doi.org/10.1007/978-981-99-3428-7_9
11. Farahmand-Tabar S, Shirgir S (2023) Incorporating nelder mead simplex as an accelerating operator to improve the performance of metaheuristics in nonlinear system identification. In: Kulkarni AJ, Gandomi AH (eds) Handbook of formal optimization. Springer, Singapore. https://doi.org/10.1007/978-981-19-8851-6_39-1
12. Farahmand-Tabar S, Ashtari P (2023) Bilinear Fuzzy Genetic algorithm and its application on the optimum design of steel structures with semi-rigid connections. In: Kulkarni AJ, Gandomi AH (eds) Handbook of formal optimization. Springer, Singapore
13. Farahmand-Tabar S, Ashtari P (2020) Simultaneous size and topology optimization of 3D outrigger-braced tall buildings with inclined belt truss us-ing genetic algorithm. Struct Design Tall Spec Build 29(13):e1776. https://doi.org/10.1002/tal.1776
14. Ashtari P, Karami R, Farahmand-Tabar S (2021) Optimum geo-metrical pattern and design of real-size diagrid structures using accel-erated fuzzy-genetic algorithm with bilinear membership function. Appl Soft Comput 110:107646. https://doi.org/10.1016/j.asoc.2021.107646
15. Elias M-P, Abdalla J (2019) Exporting water wave optimization concepts to modified simulated annealing algorithm for size optimization of truss structures with natural frequency constraints. Eng Comput 37(1):763–777. https://doi.org/10.1007/s00366-019-00854-6
16. Ou D, Zhou X, Lin M et al (2018) Singular solutions of truss size optimization for considering fundamental frequency constraints. Arch Appl Mech 89(4):649–658. https://doi.org/10.1007/s00419-018-1483-6
17. Dede T, Togan V (2015) A teaching learning-based optimization for truss structures with frequency constraints. Struct Eng Mech 53(4):833–845. https://doi.org/10.12989/sem.2015.53.4.833
18. Asadi FM, Abadi A, Cheraghi A (2022) Optimal design of truss structures with natural frequency constraints utilizing IWSA algorithm. Lect Notes Civil Eng 0(0):75–87. https://doi.org/10.1007/978-981-19-0507-0_8
19. Moosavian H, Mesbahi P, Moosavian N et al (2021) Optimal design of truss structures with frequency constraints: a comparative study of DE, IDE, LSHADE, and CMAES algorithms. Eng Comput. https://doi.org/10.1007/s00366-021-01534-0
20. Baykasoğlu A, Baykasoğlu C (2021) Weighted superposition attraction-repulsion (WSAR) algorithm for truss optimization with multiple frequency constraints. Structures 30:253–264. https://doi.org/10.1016/j.istruc.2021.01.017
21. Carvalho JG, Lemonge AC, Carvalho ÉR, Bernardino HS et al (2017) Truss optimization with multiple frequency constraints and automatic member grouping. Struct Multidiscip Optim 57(2):547-577. https://doi.org/10.1007/s00158-017-1761-x
22. Jalili S, Talatahari S (2017) Optimum design of truss structures under frequency constraints using hybrid CSS-MBLS algorithm. KSCE J Civ Eng 22(5):1840–1853. https://doi.org/10.1007/s12205-017-1407-y
23. Nguyen NT, Nguyen-Van S, Diem TT, Hoang T, Viet Dung L et al (2022) An enhanced hybrid jaya algorithm for size optimization of truss structure under frequency constraints. Adv Eng Res Appl 0(0):166–176. https://doi.org/10.1007/978-3-031-22200-9_18
24. Thanh N-V, Nga T, Nguyen-Dinh N et al (2020) Truss optimization under frequency constraints by using a combined differential evolution and Jaya algorithm. Adv Eng Res Appl 0(0):861–873. https://doi.org/10.1007/978-3-030-64719-3_95

25. Zuo W, Bai J, Li B (2014) A hybrid OC-GA approach for fast and global truss optimization with frequency constraints. Appl Soft Comput 14:528–535. https://doi.org/10.1016/j.asoc.2013.09.002
26. Anh PH (2016) Truss optimization with frequency constraints using enhanced differential evolution based on adaptive directional mutation and nearest neighbor comparison. Adv Eng Softw 102:142–154. https://doi.org/10.1016/j.advengsoft.2016.10.004
27. Ho-Huu V, Vo-Duy T, Luu-Van T, Nguyen-Thoi T et al (2016) Optimal design of truss structures with frequency constraints using improved differential evolution algorithm based on an adaptive mutation scheme. Autom Constr 68:81–94. https://doi.org/10.1016/j.autcon.2016.05.004
28. Liu S, Zhu H, Chen Z et al (2019) Frequency-constrained truss optimization using the fruit fly optimization algorithm with an adaptive vision search strategy. Eng Optim 52(5):777–797. https://doi.org/10.1080/0305215x.2019.1624738
29. Farshchin M, Camp C, Maniat M (2016) Multiclass teaching-learning-based optimization for truss design with frequency constraints. Eng Struct 106:355–369. https://doi.org/10.1016/j.engstruct.2015.10.039
30. Tejani GG, Savsani VJ, Patel VK et al (2018) Truss optimization with natural frequency bounds using improved symbiotic organisms search. Knowl-Based Syst 143(0):162–178. https://doi.org/10.1016/j.knosys.2017.12.012
31. Salajegheh F, Salajegheh E, Shojaee S (2021) Optimum design of truss structures with frequency constraints by an enhanced particle swarm optimization method with gradient directions based on emigration philosophy. Eng Optim 0(0):1–23. https://doi.org/10.1080/0305215x.2021.2011259
32. Tejani GG, Savsani VJ, Patel VK (2016) Modified subpopulation teaching-learning-based optimization for design of truss structures with natural frequency constraints. Mech Based Des Struct Mach 44(4):495–513. https://doi.org/10.1080/15397734.2015.1124023
33. Dey N, Ashour A, Bhattacharyya S (2020) Applied nature-inspired computing: algorithms and case studies. Springer, Singapore
34. Dey N (2018) Advancements in applied metaheuristic computing. In: Engineering science reference, IGI Global; Hershey, PA, USA
35. Rocha I, Parente E, Melo A (2014) A hybrid shared/distributed memory parallel genetic algorithm for optimization of laminate composites. Compos Struct 107(0):288–297. https://doi.org/10.1016/j.compstruct.2013.07.049
36. Kamyab S, Eftekhari M (2013) Using a self-adaptive neighborhood scheme with crowding replacement memory in genetic algorithm for multimodal optimization. Swarm Evol Comput 12(0):1–17. https://doi.org/10.1016/j.swevo.2013.05.002
37. Rahmi SS, Topcuoglu H (2016) A memory-based NSGA-II algorithm for dynamic multiobjective optimization problems. Appl Evol Comput 0(0):296–310. https://doi.org/10.1007/978-3-319-31153-1_20
38. Xia Z, Liu F, Gong M et al (2011) Memory based lamarckian evolutionary algorithm for job shop scheduling problem. J Softw 21(12):3082–3093. https://doi.org/10.3724/sp.j.1001.2010.03687
39. Prasad PR, Nath DK (2016) A robust memory based hybrid differential evolution for continuous optimization problem. Knowl-Based Syst 103(0):118–131. https://doi.org/10.1016/j.knosys.2016.04.004
40. Luo W, Sun J, Bu C et al (2016) Species-based Particle Swarm Optimizer enhanced by memory for dynamic optimization. Appl Soft Comput 47(0):130–140. https://doi.org/10.1016/j.asoc.2016.05.032
41. Tang D, Cai Y, Zhao J et al (2014) A quantum-behaved particle swarm optimization with memetic algorithm and memory for continuous nonlinear large scale problems. Inf Sci 289(0):162–189. https://doi.org/10.1016/j.ins.2014.08.030
42. Mavrovouniotis M, Yang S (2012) Ant colony optimization with memory-based immigrants for the dynamic vehicle routing problem. In: 2012 IEEE congress on evolutionary computation, vol 0, no 0, pp 0–0. https://doi.org/10.1109/cec.2012.6252885

43. Ranjini KSS, Murugan S (2017) Memory based Hybrid Dragonfly Algorithm for numerical optimization problems. Expert Syst Appl 83(0):63–78. https://doi.org/10.1016/j.eswa.2017.04.033
44. Han X, Liu Q, Wang L, Zhou L, Wang J et al (2018) An improved fruit fly optimization algorithm based on knowledge memory. Int J Comput Appl 42(6):558–568. https://doi.org/10.1080/1206212x.2018.1479349
45. Gupta S, Deep K (2020) A memory-based Gray Wolf Optimizer for global optimization tasks. Appl Soft Comput 93(0):106367–106367. https://doi.org/10.1016/j.asoc.2020.106367
46. Duan Q, Mao M, Duan P et al (2016) An improved artificial fish swarm algorithm optimized by particle swarm optimization algorithm with extended memory. Kybernetes 45(2):210–222. https://doi.org/10.1108/k-09-2014-0198
47. Gupta S, Deep K, Engelbrecht AP (2020) A memory guided sine cosine algorithm for global optimization. Eng Appl Artif Intell 93(0):103718–103718. https://doi.org/10.1016/j.engappai.2020.103718
48. Salam AM, Zainol AS, Ariffin K (2020) A migration-based cuttle-fish algorithm with short-term memory for optimization problems. IEEE Access 8(0):70270–70292. https://doi.org/10.1109/access.2020.2986509
49. Zong X, Liu J, Ye Z et al (2022) Whale optimization algorithm based on Levy flight and memory for static smooth path planning. Int J Mod Phys C 33(10):0–0. https://doi.org/10.1142/s0129183122501388
50. Li J, Fan C, Yi L, Qi H et al (2018) Multiobjective optimization algorithm based on kinetic-molecular theory with memory global optimization. In: 2018 13th world congress on intelligent control and automation (WCICA), vol 0, no 0, pp 0–0. https://doi.org/10.1109/wcica.2018.8630566
51. Bassel A, Jan Nordin M (2017) Mutation and memory mechanism for improving Glowworm Swarm Optimization algorithm. In: 2017 IEEE 7th annual computing and communication workshop and conference (CCWC), vol 0, no 0, pp 0–0. https://doi.org/10.1109/ccwc.2017.7868403
52. Karimzadeh Parizi M, Keynia F, Khatibi Bardsiri A (2021) OWMA: an improved self-regulatory woodpecker mating algorithm using opposition-based learning and allocation of local memory for solving optimization problems. J Intell Fuzzy Syst 40(1):919–946. https://doi.org/10.3233/jifs-201075
53. Bijari K, Zare H, Veisi H et al (2016) Memory-enriched big bang–big crunch optimization algorithm for data clustering. Neural Comput Appl 29(6):111–121. https://doi.org/10.1007/s00521-016-2528-9
54. Acan A, Ünveren A (2014) A two-stage memory powered Great Deluge algorithm for global optimization. Soft Comput 19(9):2565–2585. https://doi.org/10.1007/s00500-014-1423-5
55. Kaedi M, Ghasem-Aghaee N, Wook AC (2013) Holographic memory-based Bayesian optimization algorithm (HM-BOA) in dynamic environments. Sci China Inf Sci 56(9):1–17. https://doi.org/10.1007/s11432-013-4829-2
56. Bednarczuk EM, Jezierska A, Rutkowski KE (2018) Proximal primal–dual best approximation algorithm with memory. Comput Optim Appl 71(3):767–794. https://doi.org/10.1007/s10589-018-0031-1
57. Braik M, Al-Zoubi H, Ryalat M, Alzubi O et al (2022) Memory based hybrid crow search algorithm for solving numerical and constrained global optimization problems. Artif Intell Rev 56(1):27–99. https://doi.org/10.1007/s10462-022-10164-x
58. Yu Z, Wang A (2010) Global convergence of a nonmonotone trust region algorithm with memory for unconstrained optimization. J Math Model Algorithms 10(2):109–118. https://doi.org/10.1007/s10852-010-9143-z
59. Liu R, Jiao L, Li Y et al (2010) An immune memory clonal algorithm for numerical and combinatorial optimization. Front Comput Sci China 4(4):536–559. https://doi.org/10.1007/s11704-010-0573-6
60. Etaati B, Ghorrati Z, Mehdi EM (2022) A full-featured cooperative coevolutionary memory-based artificial immune system for dynamic optimization. Appl Soft Comput 117(10):108389–108389. https://doi.org/10.1016/j.asoc.2021.108389

61. Socha K, Dorigo M (2008) Ant colony optimization for continuous domains. Eur J Oper Res 185(3):1155–1173
62. Sadegh KM, Naseralavi S (2014) Truss optimization on shape and sizing with frequency constraints based on orthogonal multigravitational search algorithm. J Sound Vib 333(24):6349–6369. https://doi.org/10.1016/j.jsv.2014.07.027
63. Kaveh A, Ilchi Ghazaan M (2015) Hybridized optimization algorithms for design of trusses with multiple natural frequency constraints. Adv Eng Softw 79:137–147. https://doi.org/10.1016/j.advengsoft.2014.10.001

The manufacturer's authorised representative in the EU is Springer
Nature Customer Service Centre GmbH, Europaplatz 3, 69115 Heidelberg,
Germany. If you have any concerns regarding our products, please
contact ProductSafety@springernature.com

Printed and bound by CPI Group (UK) Ltd, Croydon, CR0 4YY
01/10/2026
02225900-0002